OXFORD MONOGRAPHS ON MUSIC

THE ORATORIO IN MODENA

The Oratorio in Modena

VICTOR CROWTHER

CLARENDON PRESS · OXFORD
1992

Oxford University Press, Walton Street, Oxford OX2 6DP
Oxford New York Toronto
Delhi Bombay Calcutta Madras Karachi
Petaling Jaya Singapore Hong Kong Tokyo
Nairobi Dar es Salaam Cape Town
Melbourne Auckland
and associated companies in
Berlin Ibadan

Oxford is a trade mark of Oxford University Press

Published in the United States
by Oxford University Press, New York

British Library Cataloguing in Publication Data
Data available
ISBN 0–19–816255–3

Library of Congress Cataloging in Publication Data
Crowther, Victor.
The oratorio in Modena/Victor Crowther.
— (Oxford monographs on music)
Includes bibliographical references and index.
1. Oratorio—Italy—Modena—17th century.   I. Title.   II. Series.
ML3233.C76   1992
782.23'0945'4209032—dc20
ISBN 0–19–816255–3

Typeset by Best-set Typesetter Ltd, Hong Kong
Printed in Great Britain by
Bookcraft (Bath) Ltd
Midsomer Norton, Avon

PREFACE

The bulk of modern research into the history of the Italian oratorio in the seventeenth century has centred upon Rome, the birthplace of the genre, and upon the activities of the Oratorians and Jesuits who developed and promoted it in Rome and elsewhere. In the last twenty years, however, scholars have been examining the growth of oratorio in other Italian cities in the second half of the century and have discovered that many provincial oratories (e.g. in Venice, Bologna, and Modena) had strong local connections: they responded to local religious needs, espoused local dynastic or political causes, and drew strength from local musical traditions. As a result, the common historical assumption that oratorio spread to the provinces simply through the missionary zeal of Oratorians and Jesuits now needs modifying to take account of the fact that oratorio flourished, sometimes without their intervention, wherever local conditions favoured its development. My aim in this monograph is to show how circumstances peculiar to Modena sustained a thriving oratorio tradition in the late seventeenth century.

I first visited Modena in 1971 to study the large collection of seventeenth-century oratorios housed in the Biblioteca Estense, preparatory work for a thesis on Emilian oratorio I was writing under the guidance of Denis Arnold at Nottingham University. That first visit whetted my appetite, not only for the musical treasures of the Este collection, but also for the city of Modena. Its streets, squares, museums, and churches became a happy hunting-ground during the hours of library closure. Visiting the church of the Holy Trinity where Canon Bendinelli taught the elder Bononcini counterpoint, seeing the organ that Orazio Vecchi played, and strolling by the Cinema Metropole, on the very spot where Vitali and Gianettini directed so many splendid oratorios for Duke Francesco II, brought Modena's illustrious past to life. Though the old city wall and fortress are gone, much of the Modena the Bononcinis knew survives today, a rich resource for seicento studies.

The good quality of much of the music I discovered in Modena made me long to see it revived. With the support of musical friends in the Loughborough area (Rosalind Broad, Frances Davies, John Whitworth, Howard Jones, Lyndon Gardner, Peter Lewis, John Stevenson, John Broad, Richard Howarth, and Anice and Bruce Paterson, *et al.*). I was able to mount performances of seven Modenese oratorios at Loughborough University between 1975 and 1981. In 1982, the tercentenary of Stradella's death, I was also closely involved in revivals of his music directed by Roy Goodman, Robert Meikle, Colin Timms, and Jane Glover. I gratefully acknowledge that many of the thoughts and judgements expressed in this monograph were prompted by these associations.

In the first chapter of the book I deal with general historical matters that had an effect upon oratorio in Modena: the renovation of the city and its institutions in the early seventeenth century, the development of the Cappella Ducale, the religious life of the city and court, and the political alliances which were crucial to the security and prestige of the duchy. The development of an oratorio tradition in Modena is the subject-matter of Chapters 2, 5, and 9. Here I have adopted a chronological approach, describing in three phases how the tradition developed under the management of successive court maestri. Such a strategy has inevitably led to an uneven division of the repertory (e.g. Chapter 2 covers only eleven performances, whilst Chapter 9 deals with eighty-four) but I would maintain that one can only make sense of such a large repertory by relating its contents both to the interests of its chief patron, the duke of Modena, and to the initiatives taken on his behalf by the directors of the Cappella Ducale. The imbalance of coverage is partially mitigated by the number of oratorios I have selected for detailed examination: two for the first phase (in Chapters 3–4), three for the second (in Chapters 6–8), and five for the third (in Chapters 10–14). As most of the oratorios I describe are not yet available in print, I have been liberal in quoting from both librettos and scores of the period. A summary of my findings is presented in Chapter 15.

In the process of writing the book, I have been greatly assisted and encouraged by scholars I met at research conferences in Siena (1982), Modena (1983), and Durham (1988). I am particularly indebted to Carolyn Gianturco of Rome and Eleanor McCrickard of Greensboro, North Carolina, not only for the liveliness of their correspondence over the last ten years, but also for putting me in touch with professional groups who have performed my transcriptions of oratorios in Italy and America. I am also grateful to Arnaldo Morelli of Rome for placing at my disposal many useful details about oratorio performances and librettists in Rome in the late seventeenth century. The help I received in Modena, from the librarians of the Biblioteca Estense; the Archivio di Stato; and the Accademia Nazionale di Scienze, Lettere e Arti; and from the Teatro Comunale and the Comune di Modena who jointly hosted the conference in 1983, has been invaluable. My final thanks must go to two friends in Loughborough: Niall

O'Loughlin who gave helpful advice when this book was being drafted, and Ornella Sicco who has guided me through the intricacies of Italian prosody and corrected many of my initial errors in translation from the Italian; any residual errors are my own.

I am grateful to the Royal Musical Association for permission to reprint in the present monograph (in Chapters 1, 2, 5, 6, and 9) material which first appeared in my article 'A Case-Study in the Power of the Purse: The Management of the Ducal *Cappella* in Modena in the Reign of Francesco II d'Este', *Journal of the RMA 115: 2* (1990), 207–19.

J. V. C.

CONTENTS

Abbreviations   x

Map   xi

1. Culture, Religion, and Politics in Seventeenth-Century
   Modena   1
2. 1674–1681: Benedetto Ferrari   14
3. Ferrari's *Il Sansone* (1680)   27
4. Stradella's *La Susanna* (1681)   40
5. 1682–1686: Vincenzo de Grandis and Giovanni Battista
   Vitali   58
6. De Grandis's *Il nascimento di Mosè* (1682)   69
7. Scarlatti's *S. Teodosia* (1685)   81
8. Colonna's *La profezia d'Eliseo* (1686)   95
9. 1686–1702: Antonio Gianettini   109
10. Gianettini's *La creazione de' magistrati* (1688)   121
11. Palermino's *Il Sansone* (1688)   133
12. Vitali's *Il Giona* (1689)   145
13. Pistocchi's *Il martirio di S. Adriano* (1692)   156
14. Vinacesi's *Susanna* (1694)   169
15. Conclusions   186

Appendix 1   191

Appendix 2   201

Select Bibliography   205

Index of Oratorios   209

General Index   211

ABBREVIATIONS

A-WN	Austria—Vienna, Österreichische Nationalbibliothek, Musiksammlung
GB-Lbl	Great Britain—London, British Library
F-Pc	France—Paris, Bibliothèque Nationale, Conservatoire collection
I-Bc	Italy—Bologna, Civico Museo Bibliografico Musicale
I-MOe	Italy—Modena, Biblioteca Estense
I-MOs	Italy—Modena, Archivio di Stato
I-MOs Court Registers	Camera Ducale, Registri di Bolletta de' Salariati
I-MOs MM	Archhivio Segreto Estense, Archivio per Materie, Musica e Musicisti
L	lira (Modenese)
RMA	Royal Musical Association (London)

Notes: The numbers in square parentheses following oratorio titles refer to their listing in Appendix 1.

Throughout this monograph monetary values are expressed in Modenese lire. Though the value of the lira against other currencies fluctuated from time to time, in 1688 it was exchanged at the rate of L33 = 1 Italian gold doubloon.

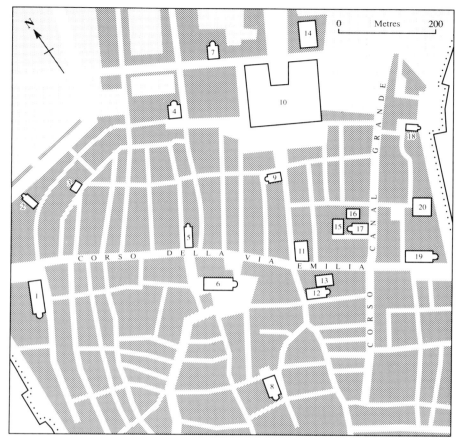

A map of central Modena showing the seventeenth-century locations of:

1. church of Sant' Agostino
2. Crocefisso church
3. oratory of SS. Annunciata
4. church of the Tertiaries
   of St Dominic
5. Voto church
6. cathedral of San Geminiano
7. Paradiso church
8. church of San Bartolomeo
9. church of San Giorgio
10. Ducal Palace
11. Fontanelli Theatre
12. church of San Carlo Borromeo
13. College of Nobles
14. Salesian convent
15. oratory of San Carlo rotondo
16. Theatine monastery
17. church of San Vincenzo
18. church of San Giovanni del
    Cantone
19. church of San Biagio
20. convent of the discalced Carmelites

I

Culture, Religion, and Politics in Seventeenth-Century Modena

The loss of the duchy of Ferrara to the papacy dealt a severe blow to the power and prestige of the Este dynasty. In January 1598 Duke Cesare d'Este was obliged to hand over his fortress in Ferrara to the papal legate, Cardinal Pietro Aldobrandini, and remove his court to the uncomfortable medieval castle in the heart of Modena. A kingdom that had stretched from the Ligurian coast to the Adriatic shrank to an area of 2,000 square miles, bordered by the Papal States (Bologna and Ferrara) in the east, the duchies of Mantua and Parma in the north and west, and Genoa and Tuscany in the south. A small retinue of Ferrarese courtiers and retainers entered Modena with their disconsolate duke on 30 January, unloaded the family treasures, and began to adjust to a mode of life far removed from the refinement and splendour they had enjoyed at the court of Alfonso II. They were faced with the task of transforming Modena from a provincial city into a capital befitting a proud and ancient dynasty.

Modena certainly had potential as a capital. It lay on the via Emilia, the main highway linking northern and central Italy, and had a network of canals supplying its citizens with fresh water and its merchants with a navigable waterway to Ferrara and the Adriatic. Produce from the *contado* (the fertile Po plain to the north and the Apennine hills to the south) was sold in the city's thriving markets or turned into fine artefacts by highly specialized craftsmen. The city was governed by the commune (Comune): a mayor (Podestà) and eight wardens (Conservatori) elected from among the nobility, acting under the auspices of the reigning duke. The outward appearance of the city had changed little since the Middle Ages. With the arrival of the court from Ferrara, however, the renovation of Modena's urban landscape became a prime objective of the duke and his subjects.

The process of renewal began in a modest way with the refitting of the castle and various dilapidated *palazzi* for the nobility, and with attempts to create a sense of unity and civic pride among the populace by arranging tournaments and open-air festivities.[1] As confidence

---

[1] See M. Calore, *Spettacoli a Modena tra '500 e '600* (Modena, 1983), ch. 3.

grew, so did the enterprise of the city's institutions. The Jesuits, ever alert to possibilities of expansion, demolished the church of San Bartolomeo in 1607 and built a new one in the Roman Baroque style with seminary attached.[2] In the same year the Tertiaries of St Dominic, a teaching order of Nuns of the Madonna, erected a church and college for thirty daughters of the gentry in the Contrada San Rocco. New churches, convents, and monasteries proliferated: the magnificent ducal church of San Vincenzo in 1617 for the Theatines, the oratory of San Carlo rotondo in 1627, also for the Theatines, the Voto church in 1634 for the Comune in thanksgiving for the cessation of the great plague of 1630, the convent of the discalced Carmelites in 1651, the church of San Carlo Borromeo in 1664, the Crocefisso church in 1668, and the Salesian convent attached to the Ducal Palace in 1670. In addition, several churches underwent major reconstruction and redecoration in the Baroque style: San Biagio in 1661, Sant'Agostino in 1662, and San Giorgio in 1680. The decision to demolish Modena's twelfth-century castle and replace it with an imposing ducal palace designed by Bartolomeo Avanzini was taken in the reign of Francesco I. Work began in 1634 and continued for over half a century.

As the physical splendour of the capital grew, so did its intellectual and cultural life. In the forefront of educational developments were the Jesuits at San Bartolomeo and the Theatines at San Vincenzo. Even before the move to San Vincenzo, the Theatines (at Madonna del Paradiso) had taken under their wing a society of pious citizens led by Father Paolo Boschetti, eager for instruction in works of Christian charity. Founded in 1608 as the Congregazione della Beata Vergine e di San Carlo, the society flourished. In 1626 it established a College of Nobles for the education of thirty young courtiers in religion, literature, science, and horsemanship. After operating in various locations, the college found a permanent site by the church of San Carlo Borromeo in 1664 and was granted university status by Francesco II in 1685.[3] The formal education of young ladies in Modena was undertaken by the Nuns of the Madonna.

Outside the major institutions of learning, citizens could find intellectual stimulation in societies and academies, the most celebrated of which was the Accademia de' Dissonanti, founded in the early 1680s by members of the court and university.[4]

---

[2] For details of 17th-cent. building projects see F. Sossaj, *Descrizione della città di Modena nell'1833* (1833; repr. Bologna, 1972).

[3] See ibid. 105–8.

[4] A fascinating account of the activities of the Dissonanti is in O. Jander, 'The Cantata in Accademia; Music for the Accademia de' Dissonanti and their Duke, Francesco II d'Este', *Rivista italiana di musicologia*, 10 (1975), 519–44.

By contrast with Bologna, where cultural affairs were governed by institutions independent of the State, Modena relied heavily upon the court of the Estensi for cultural leadership and patronage. In the seventeenth century the reigning dukes were Cesare (1598–1628), Alfonso III (1628–9), Francesco I (1629–58), Alfonso IV (1658–62), Duchess Laura (regent for her infant son (1662–74)), Francesco II (1674–94), and Rinaldo I (1694–1737). Descended from a line that had patronized writers like Ariosto and Tasso and musicians like Josquin, Willaert, and Luzzaschi, the seventeenth-century dukes continued the family tradition by bringing to the court Fulvio Testi, Velasquez, Bernini, and, in the musical sphere, Uccellini, Vitali, and Bononcini. The presence of ambassadors from France, Spain, Austria, and Rome at court kept the Modenese in touch with cultural, as well as political, happenings in foreign parts. Thus, in one way or another, the existence of a ducal court broadened horizons and enhanced the quality of life in Modena. Nowhere is that better illustrated than in the musical life of the city.

When the courtiers from Ferrara arrived in Modena at the end of the sixteenth century, the streets, inns, and squares were alive with popular songs and dances, but music of a finer sort was only to be found in the Cathedral, where Orazio Vecchi was maestro di cappella, in the larger monasteries, and in the homes of a few dilettanti. The provincial governor, Don Alessandro d'Este, half-brother of Cesare, controlled the activities of street musicians by issuing annual or occasional licences, but seemed content to leave matters of high culture to the good offices of the church. Duke Cesare brought with him from Ferrara only a handful of musicians, insufficient to constitute a ducal cappella. Spaccini's Chronicle[5] recorded the arrival in Modena of the family's precious instruments and music books. They were delivered on hand-carts in January 1601, all rain-sodden and spattered with mud, and then dumped in a store whose walls were thick with mould. To save the stringed instruments from random dispersal they were given into the care of the 'Padri' (unspecified but possibly the Theatines) who still had some respect for the happy memory of Duke Alfonso II. To be fair to Modena's new ruler, he made the best of a dismal situation. Within weeks of his arrival he recognized the outstanding talent of Vecchi and appointed him to teach music to the children of the royal household. He also invited distinguished music-

---

[5] Giovanni Battista Spaccini chronicled public events in Modena from 1588 to 1630. His handwritten observations have been partially published in E. P. Vicini, *Monumenti di Storia Patria delle provincie modenesi* (Modena, 1918), xvi–xviii. Spaccini's account of the arrival of the musical instruments is in L. F. Valdrighi, 'Cappelle, concerti e musiche estensi', *Atti e memorie della Deputazione di Storia Patria per le provincie modenese*[3], 2 (1883), 475.

ians to entertain at court: the famous cornettist Nicolo Rubino in 1618 and Sigismondo d'India in 1626. At the same time the talents of local musicians like the members of the Compagnia dei Violini, active from about 1617, were encouraged by his policy of increasing public festivities. From such beginnings as these emerged a cappella ducale, definitely established by 1629 at the beginning of the reign of Francesco I and destined to bring as much acclaim to Modena in the 1680s as the Cappella Ducale of Alfonso II had brought to Ferrara in the 1590s.

Though the size and composition of the cappella fluctuated from time to time, for much of the seventeenth century it provided stable conditions of service which attracted talented musicians to Modena.[6] Thus court, church, chamber, and theatre all benefited from high standards of performance. Table 1.1 shows the fluctuations in the size of the cappella over seven decades: stability during the reigns of Francesco I and Alfonso IV, disbandment during the regency (1662–71), revival under Francesco II until the financial crisis of 1689, and continuation under Rinaldo I until a second disbandment in 1702 when military invasion forced the court to remove to Bologna.

A clear picture of how the cappella was managed emerges from the court records of the reign of Francesco II: registers of monthly salary payments, files of bills and receipts, and items of correspondence.[7]

The system of management was autocratic. Every decision about the cappella was either initiated by the duke or required his personal consent. Once decisions were made, they were put into operation by the court treasurer Pietro Zerbini, who kept the registers, and by trusted court officials like Lodovico Tagliavini or the duke's private secretary Giovanni Battista Giardini, whose function was to liaise with the maestro di cappella over administrative details so that the duke's wishes could be carried out to perfection. With a musical duke like Francesco II at the head, the system worked well, but musicians had been alerted to the dangers of autocratic management in 1662 when his mother Duchess Laura had dismissed the whole cappella at a stroke.

Musicians were only admitted to the cappella after careful vetting, a process that could last for months or even years. They needed good references from former employers and an influential sponsor at court. Their technical capabilities were assessed by means of confidential reports from professional colleagues who knew their work, and formal auditions arranged by the maestro.

[6] For a full account of the management of the cappella see J. V. Crowther, 'A Case-Study in the Power of the Purse: The Management of the Ducal *Cappella* in Modena in the Reign of Francesco II d'Este', *Journal of the RMA*, 115. 2 (1990), 207–19.

[7] See *I-MOs* Court Registers, and MM, Buste 1–3.

TABLE 1.1. Salaried Members of the Cappella Ducale in Modena,
1629–1702

Date	Reign	Maestri	Singers	Instrumentalists	Totals
1629	Francesco I	1	6	8	15
1638	Francesco I	1	6	7	14
1659	Alfonso IV	1	5	6	12
1662	Laura (regent)	0	1	0	1
1673	Laura (regent)	0	2	6	8
1674	Francesco II	4	5	11	20
1678	Francesco II	4	9	13	26
1689	Francesco II	3	8	18	29
1691	Francesco II	3	8	6	17
1697	Rinaldo I	1	11	9	21
1702	Rinaldo I	2	10	10	22

Sources: I-MOs Court Registers, and MM, Buste 1–3.

Salaries were negotiated on an individual basis; there were no standard rates for any post. The register for 1687, for example, shows that maestro Gianettini was paid 396 Modenese lire per month, sottomaestro Vitali L128 and sottomaestro Colombi L96. Salaries of singers ranged from L198 to L32 and those of instrumentalists from L100 to L20. No doubt, in fixing the salary levels, the court took account of a musician's other sources of income. Domenico Bratti the court organist was paid a modest salary of L32 but also held the organist's post at Modena Cathedral and supplemented his earnings still further by playing the harpsichord in the theatres of Modena and Reggio. Giovanni Battista Vitali had some additional income from publishing sonatas, as did Giovanni Maria Bononcini. The latter was very poorly rewarded for playing in the court orchestra (L8 per month) but had the post of maestro at the Cathedral (L120 per annum) to eke out what his talented son Giovanni described as a miserable existence.[8]

Once musicians were admitted to the cappella their activities were carefully regulated. In return for board, lodging, and ducal protection they were expected to provide music for all court functions. The regular round consisted of church music in the chapel of San Vincenzo and chamber music in the Ducal Palace. Seasonal activities included performances of operas in the Fontanelli Theatre in Modena and the

[8] A letter dated 1683 from the younger Bononcini to his teacher Colonna, describing a poverty-stricken childhood, is cited in W. Klenz, *Giovanni Maria Bononcini* (Durham, NC, 1962), 13.

Municipal Theatre in Reggio, oratorios in San Carlo rotondo during Lent, entertainments at the duke's summer retreat, the palace at Sassuolo, in June and July, and festive music for royal birthdays and marriages. Musicians were allowed to take on contractual engagements outside the court circle but if they overstayed their leave, as the temperamental castrato Siface frequently did on his excursions to Venice, Rome, and Naples, they were severely reprimanded.

Though the main purpose of the cappella was to serve the royal household, it also gave considerable support to local Modenese institutions. Members are on record as having performed at functions in theatre, college, academy, cathedral, monastery, and church. There was enough enthusiasm for music in the city for a flourishing music-printing trade to develop in the last thirty years of the century. The presses of Ferri, Vitaliani, Cassiani, Soliani, Canori, Ricci, and Rosati found a ready market for their editions of sonatas, motets, cantatas, and music primers. Naturally, the composers and authors of most of these printed works were members of the court cappella.[9]

Francesco II's influence as a patron of music extended far beyond the court circle. The Chapter of Modena Cathedral, for example, sought his advice before appointing musicians, as also did the Comune. Further afield we find the duke securing the post of maestro di cappella at San Giovanni in Monte in Bologna for the young Giovanni Bononcini in 1687 and a similar post at the Santa Casa di Loreto for his former retainer Vincenzo de Grandis in 1685. Abroad he was recognized as a prince of excellent taste; at home his Modenese subjects were happy to follow his lead in cultural matters.

Ample evidence of the high quality and vitality of musical life in Modena can be found today in the large body of seventeeth-century compositions preserved in the Biblioteca Estense. Particularly impressive are the sonata collections[10] published over a fifty-year period—from Uccellini's Opus 2 of 1642 to Vitali's Opus 12 of 1692—and the manuscript scores of eighty-one oratorios performed between 1677 and 1702. The profusion of oratorio productions in the last quarter of the century testifies to the proficiency of Francesco II's cappella.

Though expert musicianship facilitated the growth of an oratorio tradition in Modena, the impulses that gave birth to it and shaped its subsequent development were religious and political.

The general impression gained from a study of religious affairs in

---

[9] For a list of publications see G. Roncaglia, 'Giuseppe Colombi e la vita musicale modenese', *Atti e memorie dell'Accademia di Scienze, Lettere e Arti*[5], 10 (1952), 47–52.

[10] Reviewed in E. Schenk, 'Osservazioni sulla scuola istrumentale modenese', *Atti e menorie dell'Accademia di Scienze, Letter e Arti*[5], 10 (1952), 3–28.

seventeenth-century Modena is one of a city teeming with religious institutions and charitable organizations. Among its religious personnel Modena could boast a cardinal (of the house of Este) permanently resident in Rome, a bishop and Chapter at the Cathedral of San Geminiano, monks and nuns of both Old and New Orders, professors of theology and their seminarians, court chaplains, parish priests, and lay members of religious fraternities, all with a distinctive role to play in the promotion of Christian piety.

If ambitious building projects are indicators of spiritual vitality, then the liveliest of Modena's religious institutions were the monasteries of the New Orders that had taken the lead in Catholic reform since the early years of the sixteenth century: the Theatines (founded in 1524), the Jesuits (1540), the discalced Carmelites of St Teresa (1562) and the Sisters of the Visitation of St Francis de Sales (1610). By eschewing wealth and power, and by devoting their lives to the disciplines of prayer, study, and works of charity, the New Orders had restored the morale of the church and won the respect of all sectors of society. During the seventeenth century their influence in Italy and abroad as friends of the poor and confidants of princes was profound.[11]

In tracing the early history of the oratorio in Rome, Howard Smither[12] has shown how closely its evolution was governed by the spiritual objectives of the New Orders. Both the Congregation of the Oratory, founded by St Philip Neri in 1575, and the Society of Jesus promoted the oratorio as a sophisticated vehicle for their moral pedagogy. It is not surprising, therefore, to find a similar process at work in Modena, though with the Theatines, rather than the Jesuits or Oratorians, in control. It is perhaps appropriate to take note here of the misfortune that befell the Oratorians in Modena. They had established a base at the church of Madonna del Paradiso in 1596, some two years before the arrival of Duke Cesare from Ferrara. By 1604 they had been supplanted by the Theatines and took no further part in Modenese affairs. One possible explanation of their disappearance is the fact that Pope Clement VIII (Ippolito Aldobrandini) was a close friend and disciple of St Philip Neri. When one recalls that it was Clement who forced the Estensi to relinquish their birthright in Ferrara, an act of aggression deeply resented by Cesare, the ousting of the Oratorians from Modena takes on the appearance of a political reprisal.

The change of curacy at the Paradiso church marked the first step in the rising fortunes of the Theatines in Modena. As their official

---

[11] See P. Janelle, *The Catholic Reformation* (Milwaukee, 1963), chs. 6, 10, and 11.
[12] See *A History of the Oratorio*, i (Chapel Hill, NC, 1977), chs. 1–6.

title, Order of Clerks Regular, indicates, they were a community of priests living under vows of poverty, chastity, and obedience. From their foundation in 1524 they had embraced an ascetic mode of life, engaged in pastoral work, and sought, by following their priestly vocation with utter integrity, to inspire a change of heart in all ranks of the clergy. When other New Orders sprang up in their wake, the Theatines came to be regarded as an élite corps. Ludwig Pastor has described the Order as 'not so much a seminary for priests, as at first might have been supposed, as a seminary for bishops who rendered weighty service to the cause of Catholic reform'.[13] It is easy to see why an Order of such high repute found favour with a dynasty needing to boost its self-confidence.

From their base at the Paradiso the Theatines gradually expanded their activities, first to the church of San Giovanni del Cantone and then in 1614, with the support of Prince Alfonso d'Este and his wife Isabella of Savoy, into a new convent on the Canal Grande, beside which the church of San Vincenzo was erected in 1617. The design of the church was modelled on that of Sant'Andrea della Valle in Rome. The interior decorations and statuary glorified St Gaetano (founder of the Order), the Blessed Amadeus of Savoy, and St Contardo d'Este. As curators of this splendid ducal chapel, attended regularly by all the court, the Theatines had obtained for themselves a key position in Modenese affairs.

Reference has already been made to the part played by the Order in establishing the College of Nobles in Modena. The circumstances that led to the foundation of the college also proved very important for the subsequent development of oratorio in the city.[14]

The Congregazione della Beata Vergine e di San Carlo moved with the Theatines into San Vincenzo and was granted a special altar for its regular devotions. In 1625 the fraternity split into two distinct parts: the Assumption section and the Nativity section. The former felt called to an educational role and returned to San Giovanni del Cantone to found the College of Nobles. The latter, wishing to remain faithful to the spiritual disciplines of the Rule of the Blessed Ippolito Galantini (a Florentine mystic), remained at San Vincenzo. Before long, however, some friction occured between the spiritual brothers and their hosts the Theatines. The situation was alleviated in 1627 when Prince Alfonso donated to the Nativity section three houses and a stable behind the apse of San Vincenzo, space enough to

---

[13] See *The History of the Popes*, trans. Antrobus, Kerr *et al.* (St Louis, 1891–1953), x. 418.
[14] My account of the building of San Carlo rotondo is based on G. Soli, *Chiese di Modena* (Modena, 1974), i. 223–32.

build their own oratory. The deed of gift, signed on 15 September 1627, explicitly ordered the Congregazione:

1. to erect in its oratory a tribune for the Este princes,
2. to remain under the Rule of the Blessed Ippolito Galantini,
3. to restrict membership to the laity,[15] and
4. to keep their title there and nowhere else.

The foundation stone was laid in October 1628 with Alfonso (by then duke of Modena) and all the court in attendance. By 1634 the work was finished and the Congregazione took possession of the oratory of San Carlo rotondo. The architectural plan was simple and functional: a rectangular perimeter enclosing a porch, an octagonal central space supporting a cupola (hence 'rotondo') and a high altar.[16] The main decorative features of the interior were a vision of St Charles carried by angels to heaven, painted in the centre of the cupola, and a canvas behind the high altar depicting St Charles in his cardinal's habit adoring the Nativity of Mary, above which stood the motto, 'Super populum tuum benedictio tua'. This little haven of piety, perched on the street-corner at the south-western extremity of the Theatine convent, was destined to become the main venue for oratorio performances in the reign of Francesco II.

Religious life at court was inevitably coloured by political considerations. Though the tone of religious observance was set by Theatine chaplains and Jesuit confessors, their calls for piety, asceticism, and loyalty to the pope met with some resistance when the prestige of the dynasty was at stake. Thus we find chaplains renowned for their austere mode of life being called upon to officiate at sumptuous royal weddings,[17] and confessors intensely loyal to the pope having to advise dukes who had been stripped of their dominions by the Holy Father. Conflicting loyalties were the unavoidable lot of the religious at court. The problem, of course, was not peculiar to Modena. In the Age of Absolutism any Catholic monarch whose political ambition impinged upon the vested interests of the church placed the clergy in an invidious position. It is not surprising, therefore, to find that a great many oratorio librettos of the late seventeenth century assert the supremacy of God's law over that of an ill-advised monarch.[18]

Several members of the ducal household at Modena strongly sup-

---

[15] This restriction was later waived by order of Duke Francesco I.

[16] A plan of the oratory, dated 1736, is preserved today in *I-MOs* Archivio E. C. A., F. 362. It is reproduced in Soli, *Chiese*, 229.

[17] For an account of a royal wedding in Modena see C. Oman, *Mary of Modena* (London, 1962), ch. 1.

[18] For exx. see App. nos, 4, 11, 16, 19, 22, 24, 25, 49, 62, 74, 85.

ported the work of the New Orders. Duke Alfonso III relinquished his title in 1629 in order to join the Capuchin friars.[19] His zeal for saving souls, coupled with a tendency to meddle in affairs of state, proved so embarrassing to the court that his son Francesco I was obliged to banish him to the convent of Castelnuovo di Garfagnana. A quieter mode of piety was practised by 'Madama', Matilde Bentivoglio d'Este. She provided the resources for founding a new convent for the discalced Carmelite nuns in 1651, and was entrusted by the duke with the spiritual education of the princesses. Her eldest charge, Princess Eleanora, became a nun of exemplary piety among the discalced Carmelites. The religious impulse was also strong in Laura Martinozzi.

Laura, a niece of Cardinal Mazarin, came to the court after marrying the Crown Prince Alfonso by proxy in 1655. The match had been arranged at the French court. With an astute Jesuit, Padre Andrea Garimberti, as her confessor and adviser, Laura used the twelve years of her regency (1662–74) to increase the influence of clerics at court and to ensure that her young children, Princess Maria Beatrice (1658–1718) and Duke Francesco II (1660–94), were nurtured in an environment conducive to religious devotion. Two acts of benevolence undertaken by the duchess will serve to illustrate the strength of her religious convictions and her pride in family.

In 1652, at the impressionable age of 13, Laura accompanied her mother on a visit to the convent of Salesian nuns at Aix-en-Provence. Some eight years later, as duchess of Modena, she decided to establish a branch of the Order in Modena. She asked the Congregation of Bishops in Rome for the necessary permissions and had the bishop of Modena inspect the proposed site adjacent to the north-eastern wing of the ducal palace. The first stage of the plan was accomplished in April 1669 when six nuns arrived from Aix on foot to form the nucleus of the new sisterhood. The laying of the foundation stone of the convent chapel on 17 May 1670 was celebrated with processions, singing, a mass, and cannon-fire. The Sisters of the Visitation took possession of their new house in 1672, thus fulfilling a pious intention nursed by Laura for twenty years.[20]

The other project close to her heart was the renovation of the church of Sant'Agostino, near the western gate of the city.[21] Her ambition was to create in a refurbished interior a great pantheon of the Este dynasty. The eight side-chapels and high altar were decorated

[19] See L. Chiappini, *Gli Estensi* (Milan, 1967), ch. 14.
[20] For a contemporary account of the founding of the convent see C. David, *Vive Jesus* (Aix-en-Provence, 1670); copy in *GB-Lbl* 1484 ee 4.
[21] See Sossaj, *Città di Modena*, 173–6.

with traditional images of Catholic devotion. Set in the walls of the nave between the chapels, however, were large niches containing idealized sculptural representations of legendary and historic notables of the blood royal, carved by Latanzio and Cestellino. The sequence in the nave culminated in the choir with statues of the three great Ferrarese saints of the thirteenth century, Contardo and the two Beatrices. Appropriately enough, Laura's funeral rites were conducted in this magnificent setting in 1688, but her hopes of establishing a new royal chapel at Sant'Agostino were thwarted when her son Francesco II decided to revert to San Vincenzo and the more amenable company of the Theatines.

A brighter future for the dynasty beckoned in 1673 when Laura, prompted by Louis XIV, agreed to give her daughter's hand in marriage to James Stuart, duke of York, the Catholic heir to the throne of England.[22] Princess Maria Beatrice, only 14 years old at the time, had set her mind on entering the Salesian convent but her mother, relishing the prestige of such a marriage and the opportunity it afforded a princess of the line to play a glorious part in restoring the Catholic faith to England, managed to divert her to a more challenging vocation. The wedding (by proxy) took place in Modena Cathedral on 20 September and was consummated at Dover on 21 November when the duke of York welcomed his new bride and mother-in-law to England.

The marriage had both immediate and long-term repercussions. While the duchess was away in London, Francesco cut the proverbial apron-strings and came under the sway of his ambitious and licentious cousin, Prince Cesare Ignazio d'Este. By the time Laura returned to Modena in March 1674 her son had reached his fourteenth birthday and assumed the reins of power. The new regime was not to her liking. Within two years she had packed her bags and left Modena for good to live in exile in Rome. Despite the pleas of her children to return home, she remained in exile until her death in 1687.

The long-term effect of the alliance with the Stuarts was that Francesco II was kept in a continuous state of anxiety about the well-being of his sister. He had good cause to be worried about her safety in Protestant England as almost every letter she wrote to him contained news of anti-Catholic hysteria in the realm. In a letter from London dated 4 April 1676 she reported, perhaps with some pride, that the duke of York had resolved not to attend the Protestant Church with his brother the king. She added that his resolution had aroused prejudice and caused the whole town to talk about it. By

---

[22] See Oman, *Mary of Modena*, ch. 1.

December 1678 things were much worse: her secretary Coleman had
been executed, Catholics were being forced to abandon their faith,
and she herself had had to pay off all her English Catholic servants by
order of parliament.[23]

Maria Beatrice had won many admirers in England by her charm,
benevolence, and horsemanship. But her zeal for the Catholic cause
was a constant thorn in the flesh of parliament. Charles II managed to
contain the Protestant backlash by sending the duke and duchess of
York on State business to Edinburgh and the Netherlands, but after
their succession to the throne in 1685 and the birth of a male heir in
June 1688 their fate was virtually sealed. On 10 December of that year
Maria Beatrice and her son sailed from the wharf at Whitehall Palace
into what proved to be permanent exile in France.

The impact of these events on the court at Modena was considerable. The Este–Stuart marriage was seen as a rejuvenation of the
dynasty and every dispatch from London signalling triumph (the
coronation, the defeat of the duke of Monmouth's rebellion, the preg-
nancy, etc.) or disaster (the Exclusion crisis, the exile) set tongues
wagging. On hearing of the exile of the Stuarts, Francesco II immediately petitioned Pope Innocent XI to send an expeditionary force to
restore them to the throne. The pope, who had spent most of his
pontificate trying to contain Louis XIV's schemes of aggrandizement,
politely declined the request.

For much of the seventeenth century the dukes of Modena had
pursued a foreign policy in harmony with that of the king of France.
Ties had been strengthened by intermarriage with the pro-French
families of Barberini, Farnese, Mazarin, and Stuart. Though Francesco
II was personally detested by Louis XIV for stubbornly refusing to
comply with the king's marriage plans for him, and for giving Prince
Cesare Ignazio too much executive power, Modena remained in the
French camp until Francesco's death in 1694. A change of foreign
policy occured with the accession of Duke Rinaldo. Abandoning his
cardinal's vestments, he signalled his sympathy for the imperial cause
by promptly marrying Carlotta Felicita of Brunswick. Louis had his
revenge seven years later when his army overran the duchy and
Rinaldo was obliged to beat a hasty retreat to Bologna.

As we have seen, from the humiliation of 1598 to the catastrophe of
1702 the fortunes of the Estensi fluctuated wildly. For the ordinary
citizen, however, life in seventeenth-century Modena must have
seemed quite exhilarating, with new buildings springing up in every
quarter, better opportunities for education, more public festivities,

[23] The 2 letters cited are in C. de Cavelli, *Les Derniers Stuarts à Saint-Germain-en-Laye* (Paris,
1871), i, docs. 112, 213.

and a rapid increase in cultural activities in the second half of the century. The performances of oratorios, which began in earnest in the late 1670s, were part of that cultural expansion. Though they were sponsored chiefly by the court, oratorios were accessible to citizens who were lay members of the Congregazione di San Carlo.

2
1674–1681
Benedetto Ferrari

For musicians in Modena, 1674 was an auspicious year. On 6 March Duke Francesco II celebrated his fourteenth birthday, took over the reins of power, and set about the task of enlarging the Cappella Ducale.

It had been in a parlous state for twelve years. In an economy drive, Duchess Laura had dismissed the musicians of Alfonso IV's cappella at the end of July 1662. Only one, the castrato Erculei, remained on the payroll as 'musico ecclesiastico', but at less than half his former salary. By 1671, with her son now showing some promise as a violinist under the tutelage of Giuseppe Colombi, Laura had revived the cappella on a very modest scale, retaining six instrumentalists (including Colombi) at L8 per month and two singers, Erculei and Giuseppe Paini, at L60 and L62 respectively. It was from this base that Francesco planned an expansion which by the end of 1674 would leave him with a cappella ducale of twenty musicians. As court maestro, the duke appointed a musician who had formerly served his grandfather and father, Benedetto Ferrari.

Though famous throughout Europe as a poet, composer, and virtuoso lutenist, Ferrari was not allowed to rest on his laurels at Modena. From the many letters, by and about him, preserved in the State Archive today[1] it is clear that Ferrari had to defend himself against some savage criticism at court. Only nine months after his first appointment at court[2] he wrote a long letter to Duke Francesco I complaining about the devious ways in which Padre Garimberti and the Jesuits were trying to undermine his reputation and have him removed from office. He says that it was while he was lodging with the Theatines that the Jesuits mounted their attack, a remark arousing the suspicion that there was a good deal of unbrotherly rivalry between the New Orders in Modena. Garimberti's poor opinion of Ferrari, revealed in this letter, may well have influenced Duchess

---

[1] Ferrari's letters are in *I-MOs* MM, Busta 1A.

[2] As Ferrari was appointed in Sept. 1653, the letter (undated) must have been sent to the duke in May 1654.

Laura's decision to dismiss Ferrari and the rest of the cappella in 1662; we have already noted that Garimberti was her most trusted advisor.

During the regency period, Ferrari spent an enforced retirement in his home town of Reggio in Emilia. From there in 1674 he wrote a series of letters to the court archivist Lodovico Tagliavini, expressing his hopes and disappointments during the protracted negotiations for his reinstatement as maestro di cappella to Francesco II. On 10 April he is hopeful that he will defeat his rivals Busca and Tonani in the forthcoming auditions conducted before the duke. On 30 May he reports having made a bad impression at the audition. By 2 October he is shattered by the news that he has not been restored to the post, and asks Tagliavini to prepare his final plea. The memorandum, sent by Tagliavini to the duke on the composer's behalf (undated, but presumably drafted in mid-October), is an eloquent statement of Ferrari's case. Forearmed with testimonials from an unnamed but 'most erudite' prince of the household and from Cardinal Widman, Tagliavini stresses Ferrari's great experience, international fame, and prowess on the theorbo. He excuses the composer's advanced years (70) by citing Lassus and Cavalli as shining examples of what can be accomplished in old age. The commendation ends with an apology that Ferrari will not come to court in person, being fearful that the duke might have a poor opinion of him, and thinking it best not to come and be a laughing-stock for his persecutors. We do not know exactly what transpired when the duke reviewed the case, but by 26 October Ferrari was ecstatic. Writing to Tagliavini from Reggio, he exclaims:

Thank you for the enclosed letter from S. Nardi [the duke's private secretary]. Oh, what a precious letter! Oh, what a divine revelation! How much I owe to my dear Sig^e Gio. Batta. [Giardini, a favourite of the duke in the chancellery] who took up with such speedy effect the interests of one who is his true friend and loyal servant. May God reward him.

Thus, through the advocacy of Tagliavini and Giardini, Ferrari became maestro di cappella at the ducal court, with a monthly salary of L133.

The composition of the Cappella Ducale, as reconstituted in 1674, is shown in Table 2.1, together with other changes in the membership that occured during Ferrari's tenure of office.

The recruitment drive of 1674 increased expenditure on the cappella fourfold—from L212 per month in 1673 to L979 per month in 1674. With the addition of seven more singers and a copyist in 1677, numbers rose to twenty-eight, at which level they remained stable until 1689.

TABLE 2.1. The Cappella of Duke Francesco II, 1674–1681

Name	Category	Instrument/voice	Salary in L	Yrs. of service
Ferrari, Don Benedetto	maestro		133	1674–81
Paini, Mag.[a] Giuseppe	maestro		128	1674–81
Colombi, Giuseppe	sottomaestro	violin	96	1674–81
Vitali, Giovanni Battista	sottomaestro	viola	128	1674–81
Albertini, Frate Giacinto	singer	alto	96	1674–81
Balugani, Antonio	singer	bass	32	1674–81
Bussi, Steffano	singer	?	128	1674–5
Cerlini, Vittorio	singer	?	64	1674–5
Erculei, Don Marzio	singer	soprano	60	1674
Ferretti, Basilio	singer	?	32	1675–?81[b]
Ferretti, Giovanni	singer	?	32	1675–?81[b]
Agatea, Frate Maria	singer	soprano	97	1677–81
Baraoni, Don Giovanni Maria	singer	soprano	65	1677–81
Origoni, Marco Antonio	singer	soprano	195	1677–81
Pietrogalli, Antonio (*detto* Cottino)	singer	bass	130	1677–81
Pistocchi, Francesco Antonio[c]	singer	alto	65	1677
Pistocchi, Giovanni[c]	singer	alto	65	1677
Trombetta, Agostino	singer	?	80	1677–?81[b]
Ascani, Pellegrino	instrumentalist	violone	32	1674–81
Ascani, Simone	instrumentalist	theorbo	32	1674–81

Barbieri, Don Giulio	instrumentalist	violin, violone	8–30	1674–81
Bellini, Ippolito	instrumentalist	violin, viola	8–20	1674–81
Bononcini, Giovanni Maria[d]	instrumentalist	violin	8[e]	1674–8
Bratti, Domenico	instrumentalist	organ	20–32	1674–81
Capiluppi, Francesco[d]	instrumentalist	violin, viola	8–30	1674–81
Ciocchi, Don Giovanni[d]	instrumentalist	violin	8–30	1674–81
Fermi, Don Anibal	instrumentalist	violin	8–30	1674–7
Pisani, Sigismondo Gregorio	instrumentalist	violin	32	1674–?81[b]
Severi, Pellegrino[d]	instrumentalist	violetta	8–20	1674–81
Casanova, Giovanni Battista	instrumentalist	cornetto	32	1677–81
Frignani, Antonio	instrumentalist	copyist	32	1677–81
Vitali, Tomaso Antonio	instrumentalist	violin	30	1677–81
Belletti, Mag.[a] Paolo	instrumentalist	violone	30	1678–81

[a] The Italian title 'Mag. [nífico]' indicates a courtly rank.

[b] As the court registers for the period 1679–82 are missing, the dates at which these members left Modena are uncertain.

[c] Listed in the registers until 1694 but paid only in their first year of service.

[d] These members had served the regent from 1671.

[e] Higher salaries were awarded to seven instrumentalists in 1675; only Bononcini was held at L8, possibly for having offended the duke in 1673 by defeating his favourite, Colombi, in the contest for the post of maestro at Modena Cathedral.

Sources: I-MOs Court Registers, 179, 181–3, 186.

No evidence has yet come to light to explain why the court required two maestri and two sottomaestri. The cappella of San Petronio in nearby Bologna had twenty-six members, and yet was managed by a single maestro assisted by a choirmaster. Perhaps in Modena duties were shared by common (unwritten) agreement: Ferrari taking responsibility for theatrical events, Paini for church functions, Vitali for chamber music, and Colombi for orchestral performances. Such an arrangement might explain why there was no salary differential between maestro Paini and sottomaestro Vitali. On the other hand, the existence of four senior posts in the cappella may have been simply a matter of prestige, a public display of the duke's generosity.

The entry in the 1677 court register noting the appointment of Frignani describes him as 'Ferrari's music copyist'. Whatever musical initiatives were afoot requiring the services of a full-time copyist, the power to shape their development lay with Benedetto Ferrari.

For the historian of Modenese oratorio the Biblioteca Estense's collection of manuscript scores and printed librettos of seventeenth-century provenance is an indispensable guide. There is little, however, surviving from the decade of the 1670s to furnish a clear picture of the early stages of its development. The loss of volume i of the three-volume *Raccolta d'oratorii per musica fatti cantare in diversi tempi dall'Altezza Serenissima di Francesco II, Duca di Modona, Regio, ecc. nell'oratorio di S. Carlo di Modona,* published by the ducal printer Soliani in 1689, is particularly frustrating, as it probably contained about twenty librettos dating from the 1670s.[3] In its absence we are left with only a handful of librettos printed by Cassiani and Degni, and two extant scores, on which to base an account of the genesis of oratorio in the 1670s.

The title-page of the first extant oratorio performed in Modena, *San Valeriano* [1] reads:

IL BATTESIMO | di S. Valeriano M. | Oratorio | cantato nelle Congr. della B.V. e S. Carlo | nel solennizarsi la festa | di S. Cecilia Verg. e Mart. | avvocata de musici | DA D. Marzio Erculei musico ecclesiastico di SAS | posto in musica dal Sig. Alfonso Paino | Mastro di Cappella di detta Congregazione.

The performance was evidently given on St Cecilia's feast-day, 22 November, in the oratory of San Carlo rotondo. As for the date, scholars disagree. Girolamo Tiraboschi[4] suggests 1628 (perhaps a

---

[3] Vols. ii (1680–7) and iii (1688–9) are in *I-MOe* 83-I-6 and 83-I-5, respectively.

[4] See his entry for Erculei in *Notizie de' pittori, scultori, incisori e architetti natü degli stati del Serevissimo Signor Duca di Modena, con un appendice de' professori di Musica raccolte dal cavaliere ab. G. Tiraboschi* (Modena, 1786), under 'Professori di Musica', 362 ff.

misprint for 1682) while Gino Roncaglia[5] confusingly offers both 1665 and 1682. A clue to the dating of the work lies in the author's official title. Erculei was created 'musico ecclesiastico di SAS' by Duchess Laura, and tactfully refrained from using the title under the new regime of Francesco II. Two singing primers, published by him in 1683 and 1686, carry only his name and birthplace, Otricoli. If Erculei used this title only during the regency, then Roncaglia's 1665 is the best date we have for *San Valeriano*.

Little is known about the composer, Alfonso Paino. A certain 'Mag.^{co} Alfonso Paini' appears in the 1659 court register as Foreign Usher at a salary of L28 per month—the same register lists Erculei among the singers at court—but there is no proof that the usher and the maestro of San Carlo rotondo were the same person.

The score of *San Valeriano* is lost. The libretto, however, reveals that the oratorio is a kind of dramatic ode to St Cecilia, constructed in one part. Most oratorios of the period had two, or sometimes three, parts. The urgency of the opening chorus: 'O ye faithful, run, marvel and admire the whiteness of a lily! O ye happy ones, praise its sweet perfume!', followed by the Narrator's panegyric on the virtues of Cecilia, sets the tone for a celebration of the glorious deeds of the martyrs. There are six characters in the oratorio: a Narrator, the Christian maid Cecilia, her pagan admirer Valerian, her Father who wishes the pair to marry, her Guardian Angel, and St Urban who officiates at the baptism which seals Valerian's fate. A Chorus of Angels praising God, and a Chorus of Gentiles praising Hymen, add their weight to this contest between Christian and pagan values, which ends in glorious martyrdom for Valerian and Cecilia.[6]

If *San Valeriano* was indeed produced in 1665, then a gap of twelve years ensued before further oratorios were performed in Modena. Unless further evidence comes to light filling that gap, Paino's oratorio stands as an isolated phenomenon, a brave initiative by the Congregazione di San Carlo, but one not built upon until the resources of Francesco II's cappella became available.

The earliest evidence linking the name of Duke Francesco II with oratorio comes from Bologna. On 2 March 1676 a performance of Giovanni Legrenzi's *Gli sponsali d'Ester* was given in his honour in the salon of Count Astorre Orsi, a wealthy Bolognese patrician. It seems likely that the duke's sottomaestro Vitali had a hand in the arrangement of this event, for he certainly knew the venue well, having

---

[5] See *La cappella musicale del duomo di Modena* (Florence, 1957), 299–300 nn. 1, 9.

[6] A famous series of murals illustrating the legend of St Cecilia was painted in the early 17th cent. by the school of Carracci in the courtyard of the monastery of San Michele in Bosco, Bologna.

directed his own oratorio *Agare* there in 1671 and another, *Gefte*, in 1672. The record of Count Orsi as a patron of oratorio, beginning with Cazzati's *Giuditta* in 1668, shows him to have had a great liking for the 'oratorio erotico'.[7] By choosing Legrenzi's *Ester* for this grand occasion, Orsi was both pleasing himself and, at the same time, offering his youthful and impressionable guest a delightful dish for his first taste of oratorio.

As the score of *Ester* is lost, only the Bolognese libretto remains as a record of the event.[8] Fortunately, it contains some valuable informa-tion about the nature and purpose of the performance. We learn from the vote of thanks printed on page 39 that ladies attended the salon: 'Thanks to the most eminent gentlemen and ladies who favoured us with assistance at the oratorio.' We can also gather, from Count Orsi's formal dedication of the work to Duke Francesco, that the host was using the occasion to air his own views about the duke's political and dynastic responsibilities:

The great extent of the empire of Assuero [king of the Medes and the Persians] which one hopes in our century to see reduced by the arms of Your Serene Highness to a tributary of the Este Eagle, invites a Muse, who respectfully makes play with the obscure story of that Monarch, to plead for the protection of such a great successor [as yourself]. I here present . . . the hope that Heaven will grant Your Highness the honourable destiny of setting your Catholic heel upon Asian treachery. Give the bounty of Your Highness's support to my most humble prayers, which I implore for *Gli sponsali d'Ester*, and they may be the bridesmen at the happy propagation of that most serene blood which today, in Your Highness, makes resplendent the glory of your great ancestors.

Orsi's invective against the Turks should be seen in the context of Pope Clement X's acute anxiety about Turkish incursions into Catholic Poland. The Bolognese, as citizens of the Papal States, had a duty to enlist what support they could for the Catholic cause against the invader. Unfortunately for Count Orsi, his plea fell on deaf ears: Francesco, by temperament, was a peace-loving prince, unlikely ever to lead an army to Warsaw.

The libretto indicates that Legrenzi's oratorio had choruses and instrumental pieces in plenty. The anonymous Venetian poet, feeling impelled to evoke at times the spirit of the Adriatic muses (i.e. writing

---

[7] Broadly, 'erotico' denotes oratorios that give a prominent part to beautiful women. See H. E. Smither, *A History of the Oratorio*, i (Chapel Hill, NC, 1977), 302–3.

[8] A copy is extant in *I-Bc* Sesini (ed.) Catalogue 5, item 2681. Librettos for performances of *Ester* at S. Maria della Fava in Venice (1675) and at Ferrara (1677) are also extant. see D. and E. Arnold, *The Oratorio in Venice* (RMA Monographs, 2; London, 1986), 7.

in the manner of opera librettists), sprinkles asterisks on his text. Even an innocent poem like Esther's wedding-song, 'O soavi tormenti', receives an asterisk to protect it from the ecclesiastical censor's red pencil. The festive chorus at the end of Part II, 'Si speri, si goda, si rida, sù sù', has no religious message or moral, quite unusual in an oratorio, but no doubt acceptable in the secular atmosphere of a Bolognese salon.

In the following year, an oratorio of very different character was heard in Modena. On 4 February 1677[9] the Congregazione di San Carlo convened a meeting to honour the bishop of Modena, Ettore Molza, with a performance of *S. Antonio Abbate, l'eroe trionfator dell'Inferno* [3]. The music (lost) was composed by Pacchioni,[10] a native of Modena and a former chorister at the Cathedral. The libretto by Carli contains an elaborate dedication to the bishop in which his work as a pastor and father of the Modenese flock is compared to that of St Antony[11] who conquered the demons in Hell 'with the CROSS as his sword and the adored name of JESUS as his shield'. The oratorio is divided into two parts: Primo Canto and Secondo Canto. In the first St Antony is set upon by a Chorus of Demons but, with the help of a Guardian Angel, withstands their attack and manages to repulse the blandishments of a Seductive Demon. In the second canto the devils gather for a show-down. Tigers, Dragons, Gorgons, and Ployphemes rise to their aid as Lucifer sings his incantation. Antony is saved only by the intervention of Jesus Christ, who puts the horde to flight. The oratorio concludes with a Chorus of Angels proclaiming victory and singing, 'To Antony the honour, to God the glory!'

Antonio Ferrari's oratorio, *San Contardo d'Este* [2], was also performed in Modena in 1677. The venue was probably San Carlo rotondo again, for the composer was a court chaplain and the subject of the oratorio was the most revered of the duke's saintly ancestors. Heir to the duchy of Ferrara in the thirteenth century, Contardo gave up his title to make a holy pilgrimage to the shrine of Santiago de Compostela in northern Spain. Ferrari's score carries the simple title, *Oratorio di San Contardo d'Este*. Manzini's libretto, however, has a more portentous heading, *Il senso depredato nell'abbandono del mondo dal gloriosissimo S. Contardo d'Este*, indicating that the audience is to witness the saint's assumption into Heaven. *S. Antonio Abbate* had opened

---

[9] Old Style and New Style conventions of dating were used haphazardly in Modena. Thus, any dates falling in Jan. or Feb. may have belonged to the end (Old Style) or the beginning (New Style) of a particular year.

[10] For details of Pacchioni's long career in Modena see Roncaglia, *Duomo di Modena*, 167–80, and J. A. Griffin, 'Pacchioni', *New Grove Dictionary of Music*, xiv. 43–4.

[11] The saint depicted in the oratorio is a composite, typical of the seicento, of St Antony the Hermit and St Antony of Padua.

with a loud chorus of bloodthirsty demons; *San Contardo* matches it with the blaring of celestial trumpets and the Narrator's cries of 'A battaglia!' It is the Day of Judgement, and the righteous are being called forth from the grave to be translated to Heaven. The saintly Contardo, while awaiting his summons, discusses political and family affairs with two Cavaliers who are eager to know his opinion of the Guelph and Ghibelline factions responsible for the civil wars in medieval Emilia. The poet here is using the same conversational method to throw light on Emilia's past as Dante had used in *The Divine Comedy* to interpret Tuscan affairs. At the end of Part II an Angel escorts Contardo to his rightful abode to the accompaniment of more triumphant fanfares and the acclaim of the Chorus: 'Glory to the heroic acts of the ancient Estensi who scorned worldliness and ravaged sentimentality.' Ferrari's setting of this pretentious dynastic tribute shows him to have been a pedestrian composer, familiar with Venetian practice. Though his score carries no specific instructions for trumpets to be used, he allows for their participation by setting triumphal passages in the 'natural' trumpet key of D major.

There is little extant evidence of oratorio performances in the seasons of 1678 and 1679;[12] only the score and libretto of Pacchioni's *S. Ignazio* [4] survive.

The oratorio celebrates the heroism of another staunch upholder of the faith, the martyr St Ignatius of Antioch. Pacchioni's second oratorio for San Carlo rotondo, the full title of which is *Le porpore trionfali del S. Martire Ignazio*, was sung on 1 February 1678 and dedicated to Prince Rinaldo d'Este, the duke's uncle. Rinaldo was a bachelor and ambitious to secure the cardinalate lost to his family at the death of his uncle in 1672. The dedicatory address by Ignazio Paltrineri makes oblique reference to the prince's ambition by using capital letters for 'purple'—a colour also associated with martyrdom—and by commending the strong religious character of the dedicatee. Rinaldo eventually gained his purple robes in 1686 at the request of his niece, the queen of England.

The story of Ignatius' martyrdom at the hands of the Emperor Trajan provides an opportunity to rehearse the arguments for and against the exercising of power by secular princes. The saint stands between the factions as both arbiter and victim. On the one hand, he refutes the Roman claim that Trajan is lord of the world and, on the other, restrains his own impetuous disciples from despising the world as barbarous and idolatrous. Sure of his destiny, and grasping the

[12] The score and libretto of Giardini's poetic tribute to the dynasty, *Il trionfo della fede nel martirio di S. Azzo Estense* (1678), are lost. The only evidence that the work existed is its listing among Giardini's publications in G. Tiraboschi, *Biblioteca modenese* (Modena, 1781–6), ii.

palm of Paradise, Ignatius stands trial before his enemies and is martyred. A young disciple sings a lament accompanied by two violas, and the Chorus closes the work with words of comfort and triumph.

The score of *S. Ignazio* is full of contrapuntal niceties. Pacchioni's vocal trios are modelled on the elegant trio sonatas of his violin-teacher G. M. Bononcini, and his mastery of the Roman style, learned at the feet of Canon Bendinelli at the church of Santa Maria delle Assi in Modena, is evident in the contrapuntal sinfonias and in the massive double-choruses which close Parts I and II.

Modena awoke in 1680 to the delights of the oratorio erotico: Antonio Ferrari returned to the oratory with two works, one about the attempted seduction of Joseph by Potiphar's wife, the other about King David's infatuation with Bathsheba, and Benedetto Ferrari rounded off a brilliant career with a sensuous version of the story of Samson and Delilah, dramatized by Giardini.

Antonio Ferrari's *Oratorio di Gioseppe* [5][13] is set in Potiphar's house in Egypt. One could be forgiven for thinking it a Venetian boudoir from the way the plot develops. Joseph (A.) has to resist the overtures of a lovelorn Nurse (T.) before encountering the arch-seductress Celinda (S.), Potiphar's wife. Encouraged by the Narrator (B.) he takes up arms against cupidity. His integrity impresses Celinda, who confesses to having learned from her Hebrew slave the value of honesty, justice, prudence, and obedience. In true Venetian style, Joseph and his mistress celebrate their new platonic relationship in a pair of short duets.

The sensuality of many biblical stories was a cause of anxiety for some tender consciences in the oratories. The author of Antonio Ferrari's third oratorio, *Il trionfo della penitenze, overo Il Davidde* [7],[14] defended his handling of erotic material in an address to the reader, printed in the libretto:

To the Reader.

The love between David and Bathsheba, and the penitence of both, has moved my pen to outline the events. I will not offer you the plot, knowing that it does not come to your great understanding as a new story, being one familiar even to simpletons. I will only say here that I have deleted parts of those happenings which have been considered too obscene, this a poetic ruse of mine for avoiding the accusation of being too wanton a writer by not

---

[13] The score in *I-MOe* is attributed to 'anon'. That it is definitely a composition of Antonio Ferrari we learn from the list of Duke Francesco II's music books (an 18th-cent. manuscript) repr. in E. J. Luin, 'Repertorio dei libri musicali di S. A. S. Francesco II d'Este nell'Archivio di Stato di Modena', *Bibliofilia*, 38 (1936), 418–45.

[14] The score is again attributed to 'anon'.

altering the story. Read the Sacred Scriptures if you do not believe me, and be satisfied.

I have tried to depict a king enamoured by a licentious adultress, not a continent Xenocrates or modest Penelope, therefore please excuse the freedom of my writing, and if you perceive only the harshness of what they say in the thorns, notice that among them is the rose of their exemplary penitence. . . .

Count Giovanni Battista Rosselli
1 May 1680.

Rosselli divides each part of his oratorio into three subsections. Part I begins with a brief battle-scene in which Uriah, the husband of Bathsheba, is slain. The Narrator (B.) introduces first Bathsheba (S.) and then David (A.), and leaves them to pursue their courtship in a series of arias and duets. The Chorus sounds a moral protest: 'It is a crime. Fleeting pleasure leads to Hell.' Part II begins with an ombra scene. Bathsheba is discovered raising up the spirits of Hell to ensnare the soul of David. In the nick of time the prophet Nathan (T.) makes an appearance to call the couple to repentence. The final scene is one of reconciliation and forgiveness, to which the Chorus responds in a more charitable vein, 'A river of tears cancels all errors.'

There is little to enthuse about in the musical settings of *Gioseppe* and *Davidde*. The arias, with their motto beginnings, lilting rhythms, and cadential hemiolas, are pale imitations of Cavalli or Cesti. The orchestral writing, though often in a sonorous texture of five parts, is unimaginative. Only in the concerted choruses of *Davidde* do we find invention capable of raising our spirits. Antonio Ferrari continued to practise music for some years,[15] but his career as a composer of oratorios came to an end with *Davidde*.

Though Benedetto Ferrari's *Sansone* [6] will be discussed fully in Chapter 3, it is worth noting here the connection between its subject-matter and that of the other oratorios performed in 1680. In each of them we see a seductress exposed. If they were planned to complement each other, who masterminded the scheme? There are good reasons to suspect that it was the author of *Sansone*, Giovanni Battista Giardini. He was trusted by the duke and, having noted Francesco's response to Legrenzi's *Ester* in Bologna, guessed the time was ripe to introduce the oratorio erotico in Modena. As a court official of long standing he would have been well acquainted with Maestro Ferrari and Don Antonio. Furthermore, Giardini demonstrated later in his

---

[15] Ferrari's name appears in the records of the Teatro Municipale at Reggio as first harpsichordist in a performance of P. A. Ziani's opera *Il talamo preservato* in 1683. See P. Fabbri, 'Il municipio e la corte: il teatro per musica tra Reggio e Modena nel secondo Seicento', in C. Gianturco (ed.), *Alessandro Stradella e Modena* (Modena, 1985), 175.

own literary career great skill in devising and promoting cycles of oratorios.[16] Whether Giardini was behind it or not, the fact remains that in 1680 the heroes of the Church Militant and the House of Este were replaced by biblical enchantresses as subject-matter for the devotions of the Congregazione di San Carlo.

From 1677 to 1680 the oratorio in Modena had been sustained by the efforts of local poets and composers. In what proved to be Maestro Ferrari's last season, 1681,[17] oratorios by Martini from Venice, Colonna from Bologna, and Stradella from Genoa were heard in San Carlo rotondo.

Though it is known that Giovanni Marco Martini was active in Milan in the years 1680–1,[18] his oratorio *Il trionfo dell'Anima* [9], a morality play featuring the Soul (S.), Penitence (T.), Temptation (B.), and an Angel (A.), is constructed as an abstract debate, a form common in his native city Venice. The music comprises a string of short arias and recitatives, relieved only by one duet and a concluding quartet for the solorists. Unusually, there are no sinfonias in the oratorio, even though an ensemble of four-part strings accompanies arias. Martini was capable of better work, as he later proved in chamber cantatas written for the Accademia de' Dissonanti in Modena.[19]

Gian Paolo Colonna, maestro di cappella at the basilica of San Petronio and at the church of the Oratorians, Madonna di Galliera, in Bologna, had many friends and business acquaintences in Modena: Vitali, Pacchioni, and numerous clients of his celebrated father's organ-building firm. He visited San Carlo rotondo to direct his sixth oratorio, *Il transito di San Gioseppe* [8].[20] It was Colonna's third oratorio on a text by the Bolognese poet Giacomo Antonio Bergamori, a man whose long-windedness, sentimentality, and occasional banality seem so much at odds with the sparkling professionalism of the composer that one wonders how their partnership ever survived.

The libretto of *San Gioseppe* depicts St Joseph on his deathbed among the members of the Holy Family. An earlier version of the same subject (music by Cazzati, text by Sanuti[21]), produced in Bologna in 1665, had placed the Virgin Mary at the centre of events, but Bergamori fixes the attention on Joseph (A.). Comforted by Mary

[16] Giardini's *Vita di Mosè*, a cycle of 8 oratorios, is discussed in Ch. 5.
[17] Ferrari died in post on 22 Oct. 1681.
[18] See L. Bianconi, 'Martini', New Grove xi. 725–6.
[19] See O. Jander, 'The Cantata in Accademia: Music for the Accademia de' Dissonanti and their Duke, Francesco II d'Este', *Rivista italiana di musicologia*, 10 (1975), 519–44.
[20] Other performances of the oratorio were given in Bologna (1678 and 1685) and Faenza (1693).
[21] Libretto in *I-Bc*.

(S.) and Jesus (S.), and protected by an Angel (S.) from the vindictive attacks of Lucifer (B.), the saint bids farewell to the world in a lovely bel canto aria, 'Di morte vicina, non più tormentatemi', only 8 pages before the end of a 67-page score. As the Angel sings him to Heaven, Lucifer flies off in a rage, crying 'Non sono estinto!' The battle for Joseph's soul gives Colonna the chance to compose in a variety of styles and moods. Mary's accompanied aria, 'Mesti lumi lacrimate', is a moving example of the Roman bel canto style, while Lucifer's aria con tromba, 'Al suon di mie trombe schiavatemi in campo', demonstrates the brilliance and virtuosity of the Bolognese instrumental tradition. Colonna's skill as a contrapuntist is evident in an elegant fugal trio for the Holy Family in Part I and in the Madrigal chorus in five parts which he uses (with different texts) to close Parts I and II.

As it was performed quite late in the Lenten season (16 April), Stradella's oratorio, *Susanna* [10],[22] was probably the last one to be heard under Ferrari's musical regime. The skill with which the librettist, Giardini, handles the erotic subject-matter, matching, if not surpassing, the work he did for Ferrari on *Sansone*, gives further proof that the duke's secretary in the chancellery was the man responsible for bringing the oratorio erotico to Modena. As far as we know, the composer did not attend the performance—his name is not mentioned in the libretto—so perhaps the court maestro himself had the privilege of directing Stradella's masterpiece.[23]

It appears that Benedetto Ferrari had wider interests than his reputation as a man of the theatre recognizes. He had had a passionate interest in oratorio for many years prior to the composition of *Sansone*. In a letter sent from Reggio on 1 February 1667 to Cardinal Rinaldo d'Este, he wrote:

It is the devotion of these sacred oratorios which are performed here in San Pietro, rather than rivalry, that has induced me again to compose one to the glory of God and of St Benedict, and to perform it on the last Friday of Carnival. . . . The poetry and music is my own; I shall accompany it on the Theorbo, for many of the people here are troublesome to listen to. The title is 'Il sospetto con il pianto di nostra Signore nel Viaggio del Redentore al Calvario'.

From such personal testimony as this we can see at least one of the reasons why oratorio began to flourish in Modena: it enjoyed the moral support of Maestro Ferrari.

---

[22] The oratorio is fully discussed in Ch. 4.
[23] The instrumental parts listed in A. Chiarelli, *I codici di musica della Raccolta Estense* (Florence, 1987), items 431, 437, and 438 are for vn. 1, vn. 2, and theorbo (Ferrari's instrument).

3

Ferrari's *Il Sansone* (1680)

Il Sansone, an oratorio in two parts
Music by Don Benedetto Ferrari
Libretto by Giovanni Battista Giardini

Characters:		
Dalida (Delilah)		S.
Testo (Narrator)		T.
Sansone (Samson)		B.
Ragione (Reason)		A.
Senso (Passion)		B.
Capo de Filistei (Philistine Captain)		B.
Due del Coro (Two Israelites)		A. and T.
Coro d'Israelliti (Chorus or Israelites)		S. A. T. B.
Coro a sei voci (Final Chorus)		S. S. A. T. T. B.

Instrumental accompaniment *a 5*
First performed in the oratory of San Carlo rotondo, Modena, in April 1680.

As we have seen in Chapter 2, Ferrari was nearing the end of a long and varied career when he embarked upon the composition of *Sansone*. His librettist Giardini served the court at Modena for over fifty years as an administrator, poet, and statesman. From a junior post in the chancellery of Alfonso IV he was promoted to the secretariat by Duchess Laura (1664), rose to become Francesco II's private secretary (1687), and finally was made 'Fattor della Camera' by Rinaldo I and sent to Germany to negotiate with the imperial princes over the disputed territory of Comacchio (1710). His first publications as a dramatist were an opera, *Il principe corsaro* (1674) to celebrate the accession of Francesco II, and a comedy, *Gli amori travestiti* (1675). *Sansone* was Giardini's first attempt at dramatizing a story from the Old Testament. Nine others were to follow in the next eleven years.

The Libretto

On pages 3 and 4 of the printed libretto Giardini dedicates his oratorio to Duke Francesco II in the customary flattering manner. He begins by pointing out that the eagle on the family crest of the Estensi was, in former times, the sacred emblem of Roman emperors. In the second paragraph he elaborates:

I myself have good reason to believe that, since those days, Este Virtue has been symbolized by that Eagle which, superior to all others, when feeling the effects of Cupid's flame, has always directed its flight to the stars, alone escaping the general conflagration brought on by impure emotions, and remaining ever spotless. For this reason, Samson is brought to the feet of Your Serene Highness. He is reduced to a sorry state by the fire of love. His eyes put out, his strength lost, he comes, although late, to learn in the School of a Prince, wishing now, more than ever, for the faculty of sight to make for himself, from the heart of Your Serene Highness, a mirror and pattern for controlling his own affections.

The dedication ends with the usual compliments to the patron and admissions of inadequacy on the part of the author, dated April 1680, Modena.

Giardini's drama follows the sequence of events described in the Book of Judges, beginning with the burning of the Philistine cornfields (15: 5)[1] and ending with Samson's capture and imprisonment (16: 21). The climax of the biblical account, the scene in the temple of Dagon, is omitted from the oratorio because it would have diverted attention away from the fundamental purpose of the drama, to warn against the evils of worldly love. As the final chorus expresses it, 'Worldly love saps our strength and distracts us from virtue.'

Part I of the oratorio is divided into three scenes.[2] The first celebrates the strength of Samson, Israel's champion. After a brief sinfonia the Narrator recounts Samson's heroic exploits and introduces the happy Israelites singing his glories. They attribute the salvation of Israel to the strength of his arm. Samson intervenes to remind his fellow-countrymen that Heaven alone performs miracles. The victory belongs to God. Suitably chastened, the Chorus responds with, 'Long live the God of victory!' In the second scene, Samson meets Delilah and is overwhelmed by her physical beauty. He is torn between his passion for her and his knowledge that she is an enemy, a Philistine. Delilah, angered by his indecision, resorts to crocodile tears. They have the desired effect and the hero yields, claiming that her eyes have driven him mad. For her part, Delilah scents victory and dashes off an aria in praise of Cupid, gloating that two large teardrops have been sufficient to subdue a Samson. The last scene takes the form of a moral debate in which the allegorical figures of Reason and Passion dispute for the soul of the hero. Reason exhorts

---

[1] Citations from the Bible are from the Authorized Version.
[2] I use 'scene', here and elsewhere, to indicate a coherent subsection of the text of an oratorio.

him to protect the frail flower of moral beauty within himself. Passion urges him to follow his heart. Reason compares his temptation to the evil song of the Sirens. Passion retorts with a *carpe diem*. Distraught by the arguments, Samson admits that his resistence is weakening. Finally the Narrator delivers a severe warning, 'Her smiling lips will be the death of you!' With this challenge the music ends, giving pride of place to the preacher and his thoughts on the hero's predicament.

At the beginning of Part II Delilah is persuaded by the Captain of the Philistines to deliver up Samson to his enemies. There follows a transitional scene in which Reason makes a final bid to bring the hero to his senses. Samson's agitated reply shows that the cause is lost: 'Meddling Reason, these are vain arguments . . . Fraudulent brutes are not given beautiful faces, no more than Furies are harboured in Paradise!' The crucial temptation scene begins with a love duet in which Samson takes the lead. Delilah is his sunflower, he her sun. In high delight they feel they could die for each other. Delilah then begins to question her lover about the source of his strength. Here, Giardini jettisons the biblical account of false confessions and abortive arrests to create a more intimate and convincing scenario. Samson refuses her pleas because he has been sworn to secrecy. In mounting frustration, again and again Delilah accuses him of not loving her truly, until he is driven to admit, 'My strength is in my hair.' This is the cue for a second love duet in which, significantly, Delilah takes the lead. Samson can do no more than echo her expressions of satisfaction. He has been drained of all moral fortitude. The scene ends as the Narrator describes Samson falling asleep on Delilah's lap, and Passion returns to charm him with the aria 'Sleep, unconquered Mars'.

The atmosphere of calm induced by the aria is the lull before the storm. The catastrophe looms as the Narrator reports (in recitative) the violence done to Samson by his enemies: he is shorn of his locks, blinded, and condemned to hard labour. All gushing with blood and sweat, and pouring out floods of tears, the hero delivers his final soliloquy from prison. In an impassioned recitative strewn with classical allusions to Circe (Delilah) and Sisiphus (himself) he mocks his own fate. Then, in an aria addressed to the audience, he challenges all lovers who worship Cupid to gaze on him, Samson, a living embodiment of the blind and naked god. The moral is driven home by Reason: 'You can now see more clearly because you are blind', and reinforced by the final chorus: 'Aim your affections to Heaven above, where the only true love is to be found in God. Worldly love saps our strength and distracts us from virtue.'

Giardini's version of the tragedy of Samson follows the sequence of events in the biblical account, but it is worth noting that only eleven sentences in the Italian libretto are drawn from scripture. The rest are a product of the author's imagination, working upon the conventions of seventeenth-century dramaturgy. Models for Giardini's slumber scene and love duets were plentiful in Venetian operas; moral debates and introspective soliloquies were commonplace in Jesuit dramas and Roman oratorios. Even in his resourceful use of images from classical mythology to enliven his drama, Giardini was following a trial blazed by the Roman oratorio librettist Francesco Balducci in the 1640s. With such a variety of source-materials expertly synthesized, Giardini ensured that the well-worn story was presented in a manner attractive to the literati in his audience. Among those were, of course, the Theatines who ran the oratory of San Carlo rotondo; the rules of their Order encouraged them to study literature seriously. They were also bound by a vow of chastity and charged 'to avoid talk and conversations with women, even the most upright and holy'. If Giardini wrote with them in mind, it is not surprising to find the character of Delilah, the arch-enchantress, painted in such lurid colours.

The Score

Ferrari's setting of *Sansone* calls for a cast of six singers. It was normal practice in the 1680s for so-called 'choruses' in oratorios to be sung by the soloists engaged to perform the main dramatic roles. Thus it seems likely that the Chorus of Israelites in Part I of *Sansone* was sung by Delilah (S.), Reason (A.), the Narrator (T.), and the Captain (B.), and the final chorus in Part II by Delilah, a second soprano, Reason, the Narrator, the Captain (a baritone singing Tenor 2), and Samson (B.). The role of Passion could also have fallen to the Captain. Though such complicated role-play may offend our modern notions of dramatic propriety, we should remember that audiences in the seventeenth century were supplied with a libretto as a means of identifying characters at any given moment in the performance.

The instrumental requirements are even more complicated than the vocal casting, for the laudable reason that Ferrari had very subtle notions of instrumental colouring which led him to adjust the scoring with every change of mood. The rubric 'segue con vv.' occurs frequently enough to suggest that the accompaniment is devised for an ensemble of strings: violins (treble clef), violettas (soprano clef), violas (alto and tenor clefs), and cellos and violone (bass clef). The textures vary, not only in density (continuo only, *a 3*, *a 4*, *a 5*), but also in timbre. The surface colouring for most arias, duets, and choruses is that of the violins, but the tone of the violetta is dominant in Delilah's lacrimose aria 'Infelice mio cor' and Passion's aria 'Dormi invitto Marte', and the gloomy violas are chosen for Samson's lament 'Vieni amante' in the final prison scene. In theory, the instruments required by Ferrari amount to two violins, two violettas, two alto violas, one tenor viola, and a basso continuo (perhaps consisting of cello, violone, lute, and keyboards), but in practice the versatile members of the court orchestra could be relied upon to change instruments to reduce the total complement to continuo and 4 players (single or multiple). As the duke had a dozen string-players on the payroll in 1680, the size of the instrumental ensemble was at Ferrari's discretion.

The oratorio begins with a sinfonia in D major, a curious invention of only fourteen bars duration. Scored for two violins and continuo bass, it consists of rudimentary canonic figures for the violins, in the second phase of which the cello participates (Ex. 3.1). Should this

Ex. 3.1 *(cont.)*

sketch be read programmatically? Bars 1–4, for instance, could be seen as depicting Samson's strength and impetuosity being matched by Delilah's, and bars 5–14 as prefiguring his hesitancy when faced with temptation. An even more enigmatic sinfonia occurs at the beginning of Part II. The fact that its syncopated rhythms reappear a little later in Delilah's aria 'Un vezzo, un riso', adds weight to the case that Ferrari is using the sinfonia to give his listeners a tantalizing hint of what is to follow. Certainly the movements in question are too aphoristic to merit attention on their own account.

Giardini's first scene of Israelite rejoicing is laid out like a cantata text, with choral refrains and matching ariettas for two soloists from the chorus. Ferrari supports the unitary poetic design by careful thematic and tonal organization:

A	Recitative, 'Gia dala fiamma'—Narrator	D
B	Chorus, 'Viva, viva'—Chorus	B flat
C	Arietta, 'A la chioma'—an Israelite (T.)	D minor
B	Chorus, 'Viva, viva'—Chorus	B flat
D	Recitative, 'Cessate'—Samson	D
E	Aria, 'Fù sovrana pietà'—Samson	G minor
C2	Arietta, 'A le stelle'—an Israelite (A.)	D minor
B2	Chorus, 'Viva il Dio'—Chorus	D minor

A similar ritornello structure occurs when Samson capitulates to Delilah in Part II. The scene is neatly framed by two duets. In its centre the melody of Delilah's petulant aria, 'Non è ver' che m'ami, nò', is heard five times before Samson relents:

A	Duet, 'Elitropio di speranza'—Samson, Delilah	G minor
B	Recitative, 'Ma tu, deh caro'—Delilah, Samson	G minor–B flat

CDC	Aria, 'Non è ver'—Delilah	D minor
C	Ritornello *a* 4	D minor
E	Aria, 'Che vero'—Samson	G minor
F	Recitative, 'Ma ch'io riveli'—Samson	B flat
CDC	Aria, 'Non è ver'—Delilah	D minor
G	. Recitative, 'Taci, taci'—Samson	D minor–B flat
H	Duet, 'O quanto ad ossequiarti'— Delilah, Samson	F

Typical of Ferrari's subtle handling of tonality is the dramatic shift from D minor to B flat major in the recitative 'Taci, taci', to mark the point at which the hero reluctantly unveils his secret (Ex. 3.2).

Ex. 3.2

Ta - ci, ta - ci, tu mi tor-men-ti.

Com-pia-cer-ti, com-pia-cer-ti ri - sol-vo. Sap-pi, la mia for-

- tez- za con-sis - te nel mio crin.

(Be silent, you plague me. I am resolved to humour you. Know this, my strength lies in my hair.)

The arias, duets, and choruses in *Sansone* are remarkably brief for a
work composed in 1680; rarely do they exceed twenty bars. One may
attribute this terseness to Ferrari's advanced age—his style had been
forged in the 1630s when lyricism was only just beginning to flower
in dramatic music—and perhaps also to his literary scruples. Being a
librettist himself, Ferrari treats poetry with great respect. In his hands,
the rhythmic shape of a poem is enhanced by a simple musical setting,
rather than obscured by excessive phrase-repetition or coloratura. One
obvious consequence of Ferrari's modest flights of lyricism is that the
drama as a whole develops at a lively pace.

There are 18 arias in the oratorio. In terms of formal design they
fall into three categories: through-composed (9), ternary (8), and
rondo (1). Five arias have two stanzas, and one has three. There are
three modes of accompaniment: continuo only (5), continuo with an
orchestral introduction or coda (8), and orchestral (5).

A typical continuo aria with orchestral coda is Delilah's 'Un vezzo,
un riso' which occurs in her soliloquy at the beginning of Part II (Ex.
3.3). The aria is in ternary form. In section A the broken upbeat
pattern, deriving from the metrication of lines 1 and 4, is balanced by
smoother phrasing in the setting of lines 2 and 3. Section B is marked
by a modulation to the major key and by a change to dactyllic rhythm
to emphasize the singularity of 'Una' (i.e. 'Just one kind look'). Here
too we find touches of coloratura on 'angry' and 'our death'. The coda

Ex. 3.3

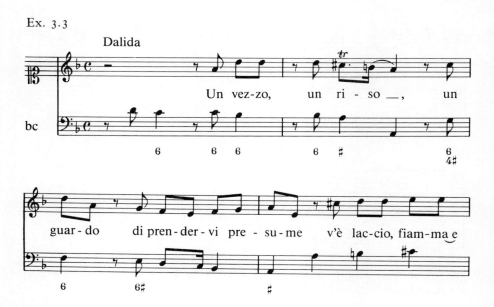

Ex. 3.3 (*cont.*)

(A charm, a smile, a glance to entice you, is for you a snare, a flame and dart, a hair, a lip, a lamp. One kind look seals our fate. One angry word shapes your terror, our death.)

for strings *a 5* is a reworking of section A, with just a hint of Delilah's
vocal line in the second violin part.

The most striking of the five arias with orchestral accompaniment is
Passion's 'Dormi, invitto Marte', lulling Samson to sleep in the arms
of Delilah. For much of the aria the bass soloist weaves his spell in
high register, wrapped in a thick cloak of chromatic harmony provided by the lower strings *a 5* (Ex. 3.4).

Ex. 3.4

violettas

vas

(Sleep, unconquered Mars. Sleep in the arms of pleasure.)

For the three duets in *Sansone* Ferrari uses the stock Venetian pattern of alternating solo statements, culminating in two-part counterpoint at the cadences. The duet which follows the hero's capitulation is dramatically effective in that Samson meekly follows (in musical imitation) every move made by Delilah (Ex. 3.5). The chorus appears only in the first and last scenes of the oratorio. It praises Samson and God in sturdy homophony at the beginning and, adding a little counterpoint to its technical repertory, closes the proceedings with a concerted madrigal.

Ex. 3.5

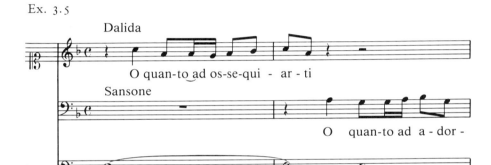

(O, how much my soul is moved to pay you homage! ‖ O, how much my soul is moved to adore you!)

Ferrari's craftsmanship is seen at its best in the handling of recitative. He follows every inflection of Giardini's text, enhancing its dramatic effect by changes of pace, accent, and mood, and by the discreet use of discord, coloratura, and arioso interpolations. At the very heart of the tragedy, recitative is the prime medium of expression. As the Narrator describes Samson's fate in the recitative 'A pena il folle amante', horror and anger give way to the pathos of the arioso 'Onde tutto grondante', with its falling sequences and poignant chromaticism (Ex. 3.6) setting the mood for the disconsolate prisoner's heart-searching soliloquy 'Da confini del mondo'. Samson's

Ex. 3.6

(. . . deprives him of sight, and condemns him to hard labour. Whence, all dripping with blood, in sweat mixed with tears . . .)

thoughts are articulated in recitative, and at some length, before he turns to challenge the audience directly in an aria.

In performance, *Sansone* lasts barely fifty minutes. Though modest in scale, it is an effective drama, set to music in a mid–Baroque manner that was virtually obsolete by 1680.

4

Stradella's *La Susanna* (1681)

La Susanna, an oratorio in two parts
Music by Alessandro Stradella
Libretto by Giovanni Battista Giardini

Characters:		
Susanna		S.
Daniele (Daniel)		S.
Testo (Narrator)		A.
Primo Giudice (Judge 1)		B.
Secondo Giudice (Judge 2)		T.
Chorus *a 3*		S. S. B. or A. T. B.
Chorus *a 5*		S. S. A. T. B.

Instrumental accompaniment *a 3*
First performed in the oratory of San Carlo rotondo, Modena, 16 April 1681.
Later performances in Bologna (1690) and Modena (1692). A facsimile of the
score (*I-MOe* Mus. F. 1137), edited by L. Callegari, is in *Bibliotheca musica
bononiensis*, sect. 4, no. 19 (Bologna, 1982).

Alessandro Stradella (b. Nepi, *c.*1639; d. Genoa, 1682) was at the
height of his powers and living in Genoa under the protection of a
friendly consortium of noblemen when his last oratorio, *Susanna*, was
composed for the Este court. He had made his reputation as an
outstanding master of oratorio composition in Rome, particularly
with *S. Giovanni Battista* in 1675, but after his hasty departure from
that city in 1677 had had little opportunity to compose further oratorios. In Genoa he was occupied with theatrical projects and occasional pieces commissioned by the nobility.

Some useful information about the genesis of *Susanna* has emerged
from letters recently discovered in the State Archives at Modena.[1]
Early in 1681 a certain Goffredo Marino, resident in Genoa, sent to
Duke Francesco a copy of Stradella's opera, *Il trespolo tutore*, delivered
by Marc'Antonio Origoni, a castrato in the Modenese cappella. In
the letter accompanying the score, Marino praised its exceptional
dramatic qualities and suggested that, if the duke were pleased with it,
the composer might be persuaded to visit Modena. Exactly what
transpired after the duke received the gift and the letter is uncertain—

---

[1] Reported in C. Gianturco, 'Il trespolo tutore di Stradella e di Pasquini: due diversi
concezioni dell'opera comica', *Venezia e il melodramma nel Settecento* (Florence, 1978), 190.

no official reply is extant—but within three months Giardini's libretto of *Susanna* had been set to music by Stradella and was being performed in Modena. It is most unlikely that the composer, who was virtually a refugee in Genoa, attended the première as his name was not published in the libretto. Later in the year, however, other reports from Genoa addressed to Giardini[2] touch upon Stradella's activities, suggesting that a good rapport had been established between the composer and librettist during their collaboration over *Susanna*.

The Libretto

Giardini's choice of the story of Susanna and the Elders was nicely calculated to suit Stradella who, in life as in art, had a keen eye for beautiful women. For the oratorios of Rome Stradella had already produced vivid portraits of Esther, Eudoxia, Edith, Pelagia, and Herodias' daughter (Salome); for the duke of Modena he was invited to add the lovely Susanna to the pantheon. He evidently relished the chance to create another great castrato role and readily took the bait.

The author's Dedication, dated 16 April 1681, touches upon the central ethical question raised in the oratorio: how can chastity be defended when officialdom is corrupt?

Susanna, a fugitive from the chaos of Babylon, having been rescued from the power and interrogations of those wicked Ministers, seeks refuge in the most happy realms of Your Serene Highness, certain that her honour will at last be accepted as free from calumnies, and her chastity safe from any sordid crime. She now carries to the feet of Your Serene Highness my poetic tribute, obscure and weak in itself until it can gain light and vigour from your most kind protection and sympathetic appreciation

The oratorio is based on the History of Susanna in the Old Testament (Vulgate version). Because the source is relatively short and self-contained, Giardini makes more use of it in his adaptation than, for example, he had done the previous year in reworking the Book of Judges for *Sansone*. His general strategy in constructing *Susanna* is to divide the story into sections of narrative (including a moral commentary), interspersed with dramatic scenes in direct speech for the protagonists. References to the biblical source are as numerous in the dramatic scenes as they are in the narrative sections of the oratorio:

---

[2] See M. Lucchi, 'Stradella e i duchi d'Este: note in margine a documenti d'archivio e agli inventari estensi', in C. Gianturco (ed.), *Alessandro Stradella e Modena* (Modena, 1985), 107–15.

Sinfonia
Part I:

Narration 1	Narrator, Chorus: Babylonian lust invades the hearts of two Israelite judges (History of Susanna 1, 5, 8, 11, 14).
Scene 1	Judges: The old men confess their cupidity.
Narration 2	Narrator, Chorus: Susanna enters her garden and bathes (History of Susanna 15).
Scene 2	Susanna: She rejoices in nature and God's goodness.
Narration 3	Narrator: The Judges prepare to seduce Susanna.
Scene 3	Susanna, Judges: She resists their overtures and threats (History of Susanna 20–4, 27).
Narration 4	Narrator, Chorus: Susanna is imprisoned; moral.

Part II:

Narration 5	Narrator: Women, beware beauty; Susanna weeps in prison.
Scene 4	Susanna: She bewails her beauty.
Narration 6	Narrator: At dawn Susanna is brought to trial. (History of Susanna 28).
Scene 5	Judges, Narrator, Susanna, Daniel: The trial of Susanna; the Judges are exposed by Daniel (History of Susanna 35–59).
Narration 7	Narrator: An attack on legal corruption.
Scene 6	Judges, Susanna, Narrator, Chorus: The Judges die in misery but Susanna is exonerated and praised by all; moral (History of Susanna 63).

The Narrator is the crucial figure in the oratorio. His voice is that of the author, all-seeing, wise, and persuasive. As his eloquence prepares the ground for each dramatic scene, so his moral judgements condition our responses. Here is no minor functionary, no simple storyteller. He is a master of ceremonies, setting the tone throughout the drama. His hyperbolic preface to Susanna's soliloquy in Part I abounds in the conceits of *marinismo*:

> Giunta la donna ove svenato un'sasso
> In conca d'alabastro
> Spande lubrico argento,
> Dove frondoso cerro,
> Briareo vegetante,
> Con cento braccia e cento
> L'ingresso al sol contende
> E da curiosi rai,

Mantenitor dell'ombra, il rio difende.
Ivi tuffa nell'acque il petto ignudo
E, sirena del Ciel,
Dentro il liquido gel così confonde
Crome di foco a l'armonia dell'onde.

(The woman having arrived where a cleft stone gushes forth liquid silver into a shell of alabaster, where a leafy oak, a Briarian plant, bars with a hundred branches entrance to the sun and, as keeper of the shade, defends the brook from prying sunbeams, there she plunges her naked breast into the waters and, like a siren from Heaven in the icy liquid, mingles the fiery colour of her skin in the harmony of the waves.)

And here, in sardonic mood, he reflects on the judicial process (Narration 7):

Promulgato il decreto,
Si corre ad eseguir.
Con tal prestezza
Nel pretorio del ciel vanno i processi;
Non così nella terra
Ove, stancati e smunti,
Egualmente puniti
I rei sono e gl'attori;
Ove a le liti si
Da vita immortale
Col balsamo dell'oro,
E con gola venale
Le sostanze contese assorbe il foro.

(Once the decree is promulgated, they rush to execute it. With the same speed actions are taken in the court of heaven; quite unlike on earth where, bored and exhausted, petitioners and the guilty are equally punished, where immortal life is given to disputes by the elixir of gold, and the court sucks contested property down its greedy throat.)

In interpreting events, the Narrator gets staunch moral support from the Chorus. Its dictum at the close of Part I prophesies the downfall of the Judges: 'Impure love has always been lethal', whilst that at the end of Part II stands as a severe warning to the audience: 'Whoever fires malicious arrows at the innocent should always expect heavenly arrows in return.'

Having injected a strong moral tone into the narrative sections of his oratorio, Giardini treats the dramatic episodes more freely, almost as though he were writing an opera for the Venetian stage. The salacious Judges are thinly disguised buffoons. The prophet Daniel makes an entry in Scene 5 as impressive as any deus ex machina:

Dove, dove correte,

> Popoli sconsigliati?
> Fermate, sospendete!
> Al sommo trono Susanna appella.
> Daniele io sono,
> Dal'eterna sapienza
> Giudice delegato
> A conoscer la colpa e l'innocenza.

(Where are you running, deluded people? Hold, stop! Susanna appeals to the highest authority. I am Daniel, a judge delegated by the Eternal Wisdom to discern guilt and innocence.)

The only rounded character in the oratorio is the heroine Susanna. At her first appearance in Scene 2, her image is that of a delightful nymph. Under stress in Scene 3, however, she is instantly transformed into a chaste wife, vigorously protesting:

> Così d'Helchia la figlia
> Da voi si vilipende!
> Così da voi s'offende
> Di Gioachin la moglie!
> Ministri senz'honore,
> Giudici senza senno,
> Fuggite, fuggite indegni!

(Thus the daughter of Chelcias is despised by you! Thus the wife of Joacim is offended by you! Dishonourable ministers, judges without wisdom, away, worthless men!)

In Part II, she has to struggle through a guilt complex in a condemned cell and face the trauma of a public trial before experiencing the relief and joy of acquittal and triumph.

Giardini's *Susanna* is a bold dramatization of an erotic Bible story and, as such, comes close at times to being voyeuristic. In order to forestall rejection by the ecclesiastical censors, the author had the good sense to ensure that quotations (in Latin) of relevant verses from the Vulgate were printed in the margin of his libretto as proof of his orthodoxy and honest intentions. Had he not been a favourite of the duke, however, one wonders whether he would have got away with it.

The Score

If Giardini's dealings with Stradella over *Susanna* began when *Il trespolo tutore* was sent to the duke, then the score must have been completed in some haste, perhaps within twelve weeks. That could explain why the composer opted for a simple accompaniment for strings *a 3*, and why the Sinfonia in C major which precedes Part I is identical to one of his trio sonatas,[3] an undated piece, but perhaps one that lay conveniently to hand when the oratorio was commissioned. Given the speed at which Stradella had to work, it is remarkable that he produced music of such high quality throughout.

Though the oratorio includes choruses, it can be performed using five solo voices. Indeed, it seems to have been scored with that in mind: the moralizing choruses at the end of Parts I and II bring together all the soloists in quintet, and the singers involved in the three-part choruses are characters in the drama sympathetic to the sentiments expressed in the text. There is therefore little confusion of role-play when *Susanna* is performed without an independent chorus.

From an extant testimonial in the State Archive (*I-MOs* MM, Busta 1B), dated 9 May 1681, it appears that Giuseppe Donati (Il Tintorino) was in Modena at the time of the performance of *Susanna*, and probably also Francesco Grossi (Siface), whose name is listed with Donati's in the 1683 register. As these two famous castratos had contributed to the resounding success of Stradella's *S. Giovanni Battista* in Rome six years earlier, it seems likely that they were engaged to play the parts of Susanna and Daniel in Modena in 1681. A technical appraisal of the first and second soprano roles in these two oratorios by Stradella reveals that the singer playing Salome/Susanna requires cavernous lungs and a highly sensuous top register, while the singer playing Herodiade/Daniel requires forceful attack and great agility in middle register. The similarities are so striking that they transform a likelihood into a probability: that the soprano parts in *Susanna* were tailor-made for Donati and Grossi.

The Sinfonia which introduces *Susanna* is in four movements: the first and third are through-composed in common time, the second and last, binary in 6/8 time. It was a fitting choice for an oratorio destined for Modena because it demonstrated Stradella's mastery of fugal technique, a skill much admired in the city of Bononcini and Vitali. With the exception of the ceremonious opening phrases in free invertible counterpoint, the Sinfonia consists of four fugues of great vitality. The most ambitious, the third, is sustained for forty bars, with sixteen entries of the subject and sprightly episodes derived from the countersubject.

There are no thematic links between the Sinfonia and the music which follows it, but its key, C major, proves to be an important connection, in that Stradella chooses the key with no sharps or flats to symbolize innocence or virtue in the oratorio. Thus the tonality of the

---

[3] Printed in E. McCrickard (ed.), *Concentus Musicus* (Cologne, 1980), v, item 13.

Sinfonia is sustained in Narration 1 to the point at which the Chorus celebrates Susanna's virtue. As the focus changes to the lechery of the Judges the music moves gradually into flat keys. Susanna makes her entry in a sunny D major, but at the moment she is threatened by evil, Stradella reverts to C major for her protestations of injured innocence. In the trial scene in Part II, the heroine's confession and prayer, and the dramatic arrival of Daniel are cast in C major, as are the two choruses which hymn the triumph of virtue at the end of the oratorio. These recurrences are not accidential; in other large dramatic pieces like *S. Giovanni Battista* or *Il barcheggio* (Genoa, 1681) we find Stradella organizing his musical ideas within the framework of a rational tonal scheme.

The arias and ensembles in Stradella's last oratorio are of ample dimensions, the arias running to an average length of sixty bars, the ensembles to forty-three. Textures ranging from *a 2* to *a 5* are integrated by the consistent use of imitative counterpoint.

The arias are clearly set apart from their surroundings; there is none of that Middle Baroque habit of slipping casually from recitative into aria here. Eleven arias are heralded by the violins, two more by the laying-out of ground basses, and a further two by lively continuo introductions. The only aria that springs from nowhere, as it were, is one shared by the Judges in the garden scene when they are urging the aggrieved Susanna to cast care aside. The distribution of arias among the five soloists accords with their dramatic importance: Susanna has 6, Daniel 3, the Narrator 3, the first Judge 2, the second Judge 1, and the remaining aria is shared by the Judges. In terms of formal design, the arias fall into three patterns: ABB (8), ABA (4), and through-composed (4). There are three modes of accompaniment: continuo only (3), continuo with ritornello(s) (7), and orchestral (6). Typical of Stradella are the free-wheeling ostinato basses heard in four arias. Only in the heroine's 'Da chi spero aita o Cieli' does the composer attempt anything approaching a regular set of variations on a ground. The result,[4] a beautiful fantasia *a 4* for voice and strings, is one of the most poignant laments of the seventeenth century.

The most arresting of the continuo arias in the oratorio is Daniel's 'Così và, turbe insane'. The young prophet castigates the crowd for giving way to blind passion in its judgement of Susanna. The frenzy of the Israelites is depicted in brilliant figurations for the continuo, the sternness of Daniel in measured anapaests and messe di voce. His mounting anger explodes in a cascade of semiquavers at 'sol regge l'empietà' (Ex. 4.1).

---

[4] Printed in A. Schering (ed.), *Geschichte der Musik in Beispielen* (Leipzig, 1931), no. 230.

Ex. 4.1

Daniele 20

Quan-do cie-ca pas - si-o-ne,quan-do cie-ca pas-si - o -ne tur-ba

bc

l'u-so à la ra- gio-ne, La Gius - ti - zia và sban-di - ta, và la

leg - ge pre -ver -ti - ta, e sol reg-ge l'em -pie-tà

25

(When blind passion disturbs the use of reason, justice is broken up, the law is perverted, and impiety rules alone.)

In the garden scene in Part I, Susanna sings three charming arias, praising in turn the brook, the fountains, and the summer breezes. The last of these, 'Zeffiretti che spiegate', is accompanied throughout by two violins (Ex. 4.2). The music, in the best traditions of Arcadia, is full of picturesque details. The dainty interplay of the violins in the introduction suggests the play of the zephyrs in the branches. Thereafter, the aspirated fanfare motif at 'sù, sù, spirate', the running figure at 'vagar', the broken gusts on 'au_re', and murmuring measured trills all imitate the various moods of summer breezes in Italian gardens. The attribution of these delights to 'God alone' occasions spondaic rhythms supported by the strongest of all tonal progressions: IV – V – I.

The four ensembles and five choruses in *Susanna* are accompanied only by the continuo instruments, though most have ritornellos attached. The ensembles are show-cases for the Judges, requiring great dexterity and rhythmic control. The sheer flippancy of their first

Ex. 4.2

Ex. 4.2 *(cont.)*

Ex. 4.2 *(cont.)*

(Little zephyrs who play here, refreshing the plumes . . . Up, up! Breezes, breathe softly, never cease to wander)

duet, 'Chi dama non ama, Villano si fà', stimulates Stradella to toss off one of his most bizarre studies in contrapuntal syncopation (Ex. 4.3). Two other duets in a lively parlando style follow, characterizing the Judges as buffoons. After exposure at the trial, however, their mood changes. In their last ensemble, a trio with Susanna, their expressions of grief and foreboding are combined with Susanna's expressions of relief and joy. Where they woefully lead, she cheerfully follows (in imitation but with a transformed text) (Ex. 4.4). Note that the return of Susanna's confidence in God is underlined by a return to the sunny D major key of her garden soliloquy.

The choruses offer a moral commentary upon the dramatic events and thus occur only in sections of narration. The style of the settings is madrigalesque, intricate counterpoint occasionally giving way

Ex. 4.3

Ex. 4.3 *(cont.)*

(He who resists Love has no feeling. No, no, has no feeling.)

Ex. 4.4

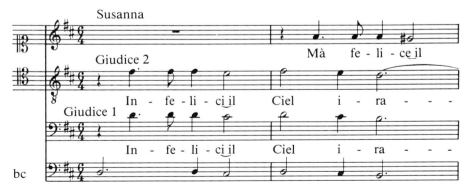

Ex. 4.4 *(cont.)*

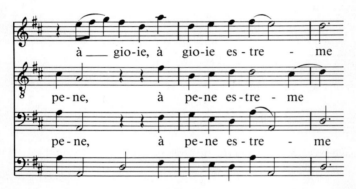

(Unhappily, angry Heaven keeps us, keeps us for punishment, for extreme punishment. ‖ But happily, placated Heaven lifts me up to joy, to extreme joy.)

to homophony when important phrases in the text call for special emphasis. In general, the three-part choruses demand as much vocal agility and co-ordination as the ensembles, whereas the five-part choruses which conclude Parts I and II achieve their effects by contrasts of density and register (Ex. 4.5).

Recitative is the chief vehicle of expression for dialogue, narration, and soliloquy. Giardini's libretto contains very little dramatic dialogue; the only extensive passage occurs in Scene 3 where Susanna is defending her honour (Ex. 4.6). The Jewess acquits herself bravely with high-pitched threats of divine retribution. Rapid key changes,

(The deadly flame is a poison and plague of the heart.)

Ex. 4.6

Ex. 4.6 (*cont.*)

-pa - ra   dell' i - ra   nos-tra à più te-mu - ti scem - pi.

15 Susanna   Giudice 2

Ben sà Dio rin-tuz-zar   l'ar-mi de-gl'em - pi__ .   Ac -

-cu - se,   in-gan - ni,   e   fro - di   ad-om-bre-ran-no la tua

                           Susanna   20

vi - ta, il tuo ho - no - re.   Va - ne min-ac - cie   Ah'__

Ex. 4.6 *(cont.)*

l'in-no-cen-za im-bel-le as-sis - te il Cie - lo,

(Yield, o beautiful woman, your love, or prepare yourself for a more
dreadful ruin from our anger. ‖ It is well known that God blunts the
weapons of the wicked. ‖ Accusations, tricks and cheating darken your life
and honour. ‖ Vain threats! Ah! Heaven assists timid innocence.)

amounting to nine modulations in the space of twenty-six bars, keep
the scene on the boil.

In narration and soliloquy, Giardini writes mainly in unrhymed
verse, but his habit of rounding off a speech with a rhyme invites
arioso embellishment in the musical setting. This can be seen in the
Narrator's account of Susanna bathing in the brook, the recitative
'Giunta la donna'. The first paragraph describes the garden. Stradella
evokes a peaceful mood with long pedal-points in the basso continuo,
gently moving into arioso in the last three bars. A surprising
shift of key from C major to E major draws attention to the figure
of Susanna, swimming like a heavenly siren in the water. The final
phrase of Giardini's text, 'Crome di foco a l'armonia dell'onde',
is reiterated with growing intensity and delight in the arioso style
(Ex. 4.7).

Ex. 4.7

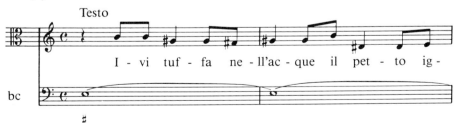

Testo

I - vi tuf - fa ne - ll'ac - que il pet - to ig -

bc

Ex. 4.7 *(cont.)*

(There she plunges her naked breast into the waters and, like a siren from Heaven in the icy liquid, mingles the fiery colour of her skin in the harmony of the waves.)

The dedicatee of *Susanna* was evidently impressed with the oratorio. After Stradella's assassination in February 1682, the duke began to build up a collection of his music[5] and took pleasure in revivals of his operas and oratorios at court. As for Giardini, his confidence as a librettist knew no bounds. After *Susanna*, he embarked upon an ambitious cycle of eight oratorios on the life of Moses, a project which was to keep him occupied until 1691.

---

[5] The most recent research papers on the Stradella collection in Modena are in Gianturco (ed.), *Alessandro Stradella e Modena*, 99–135.

5

1682–1686
Vincenzo de Grandis and
Giovanni Battista Vitali

By the end of 1681 the Cappella Ducale had lost both its senior members: Ferrari had been buried at the Paradiso church in Modena, and Paini had either resigned or been dismissed for overspending his allowance.[1] In January 1682 Don Vincenzo de Grandis, formerly maestro at the Gesù in Rome and at the court of Hanover, became head of the ducal cappella. The surviving notice of his appointment is undated, but a letter from one of his Roman referees, dated 17 January,[2] would indicate that he had taken up the post before Lent. De Grandis remained in Modena for only sixteen months. A note of good service, signed by the duke in April 1683, states that he had to leave Modena to attend to urgent family matters at his home town of Montalboddo. He stayed on good terms with the duke, however, and through Francesco's influence obtained the post of maestro at the shrine of the Santa Casa di Loreto in 1685. From there, he kept up a lively correspondence with Giardini, bewailing his misfortunes, asking about Modenese affairs, gossiping about famous musicians or ruthless cardinals, and praying for the confusion of all heretics.

After de Grandis left the court, his duties were undertaken by Sottomaestro Vitali. Though he took on the responsibility without enhancement of official status or pay, his commitment to the task was exemplary. During the three years of his caretaker-management, Vitali published four volumes of instrumental music and a collection of sacred hymns for solo voice and instruments, composed nine chamber cantatas for the Accademia de' Dissonanti and an oratorio, *L'ambitione debellata*, in honour of the queen of England, and directed or managed performances of 17 oratorios in San Carlo rotondo.

The only surviving registers of the de Grandis–Vitali era are for the years 1683 (only the index of names, no payment details) and 1684. They show a few changes in the membership of the cappella. Four

---

[1] See G. Roncaglia, *La cappella musicale del duomo di Modena* (Florence, 1957), 299.
[2] The documents and letters relating to de Grandis are in *I-MOs* MM, Busta 1A.

singers left the court: Cerlini, the Ferretti brothers, and Trombetta. Two of their replacements were castratos of international fame, Grossi and Donati. Giovanni Francesco Grossi *detto* Siface came to Modena in 1679 and served at a monthly salary of L198 until his assassination in 1697. Giuseppe Donati *detto* Il Tintorino o Gioseppino di Baviera was certainly in Modena in 1681 as the duke gave him a testimonial on 9 May. His name, with that of Grossi, appears in the 1683 index, but he must have left Modena in that year as his name is not listed in the 1684 register. Grossi and Donati were professional colleagues of long standing. They had sung together in oratorios in Rome in 1675,[3] and appeared together on the operatic stage in Venice in 1680.[4] Their likely involvement in Stradella's *Susanna* has been discussed in the previous chapter.

Two other singers were engaged in 1684: Nicolo Grancini (at a salary of L50) who left in 1686, and Giovanni Marovaldi (L99) who remained on the payroll until 1690. In the court orchestra, the violinist Pisani was replaced by Antonio Allemani, who served at a salary of L100 until 1702, while two other instrumentalists, Francesco Giberti and Gioffredo Passo (lute), served for short periods under Vitali.

As far as we know, only two oratorios were produced in Modena while de Grandis was maestro: his own setting of Giardini's *Il nascimento di Mosè* [11] and Pacchioni's *La gran' Matilde d'Este* [12], both in 1682. The score of a third oratorio, de Grandis' *La caduta d'Adamo*, found its way into the ducal library at this time, but was probably never performed. No libretto is extant, no duplicate score or parts are mentioned in the library catalogue,[5] and the author, Cavalier Nencini, was a Roman poet active while de Grandis was employed in Rome in the early 1670s. The score, beautifully copied and bound, has every appearance of having been a gift from a grateful maestro to his new employer.

Having dedicated his first oratorio to a bishop and his second to a prince, Pacchioni dedicated his third and last to Duke Francesco. It celebrates Matilda of Tuscany (1046–1115), one of the duke's ancestors and the foundress of Modena Cathedral. She was a mystic and a disciple of Pope Gregory VII, who sent her on a diplomatic mission to Canossa in 1077 to plead the papal cause in the debate on the investiture of the Emperor Henry IV. This episode in her life, together with her marriage to Guelph V of Bavaria, is the basis of the

---

[3] See R. Casimiri, 'Oratorii del Masini, Bernabei, Melani, Di Pio, Pasquini e Stradella', *Note d'archivio per la storia musicale*, 13 (1936), 157–69.

[4] See S. T. Worsthorne, *Venetian Opera in the Seventeenth Century* (Oxford, 1953), 170–1.

[5] See E. J. Luin, 'Repertorio dei libri musicali', di S. A. S. Francerco II d'Este rell'archivio di stato di Modena', *Bibliofilia*, 38 (1936), 418–45.

oratorio. The text is an untidy mixture of political and amorous intrigue, which Pacchioni furnishes with a string of uninspiring recitatives and arias. The Chorus ends the work with the predictable sentiment 'the fame of the Estensi resounds for evermore.'

De Grandis's oratorio on the birth of Moses was presented at San Carlo rotondo in May 1682.[6] It marked the beginning, at least for the author Giardini, of a huge project: to depict the life of Moses in a cycle of eight oratorios. It took Giardini nine years to write the texts, commission the music, and see his cycle performed before the duke and the Congregazione di San Carlo. The composers who contributed to this unique enterprise were:

> de Grandis, *Il nascimento di Mosè* (1682); *Il matrimonio di Mosè* (1684)
> Colonna, *Mosè legato di Dio* (1686)
> Perti, *Mosè conduttore del popolo ebreo* (1685)[7]
> Pasquini, *I fatti di Mosè nel deserto* (1687)
> Gianettini, *La creatione de' Magistrati* (1688); *Dio sul Sinai* (1691)
> Melani, *Lo scisma del sacerdozio* (1691)

The idea of a cycle of oratorios was not new to northern Italy. In Bologna in 1661 Arresti had produced a pair of matching oratorios for performance on Holy Thursday and Good Friday. Also in Bologna, Cossoni and Cazzati, aided by the enterprising poet Archdeacon Savaro di Mileto, had produced an *Adam–Cain–Flood* sequence in 1663–4.[8] We have already observed in Chapter 2 that the Modenese explored the common theme of biblical enchantresses in the 1680 season. But if the idea was not new, the sheer scale and dramatic consistency of Giardini's *La vita di Mosè* makes it a unique achievement. The author knew his duke, his audience, and his Bible, and therefore fashioned the story of Moses into a sophisticated blend of religion, politics, and dramatic entertainment that would give both instruction and pleasure.

De Grandis completed his setting of *Il matrimonio di Mosè* [15] at Montalboddo on 5 February 1684, posting it to the duke on the 26th. In a covering note, he thanked the duke for agreeing that he should set Giardini's text to music, and explained that the work was undertaken at a time when he was immersed in sorrow through the death of his only son, and other domestic anxieties.

For the performance of the oratorio in Modena, Giardini adopted a longer title, *La ritirata di Mosè dalla corte d'Egitto e suoi sponsali con*

---

[6] For details of the oratorio, see Ch. 6.

[7] The episodes set by Colonna and Perti were performed, it seems, in reverse order.

[8] On oratorio cycles see J. V. Crowther, 'The Development of Oratorio in Emilia, 1650–1700', Ph.D. thesis (Nottingham, 1977), 77 ff., 115 ff.

Sefora. The plot covers Moses' flight from Egypt into Midian, his defence of Zipporah and her sisters at the well, and his meeting with Jethro and his wife, all in Part I. The courtship and nuptuals, culminating in a choral prayer to Hymen to bless the marriage, occupy the whole of Part II. With such a rambling plot involving many characters, Giardini makes frequent use of a Narrator (Bar.) to explain the events, thus allowing him to reduce the active participants to four: Moses (A.), Zipporah (S.), Jethro (T.), and Jethro's wife (S.). The young Moses is depicted as a courtier of exemplary gallantry; 'the Cavalier who defended me' is Zipporah's description of him. Perhaps that is why de Grandis evokes so often in his score the music of the French court, a style familiar to the Modenese through the sonatas of Vitali and Colombi. The central section of the sinfonia to Part I is actually entitled 'Canzona alla Francese', and the chorus at the end of Part I trips along in the manner of a French minuet.

In 1684, under Vitali's management, San Carlo rotondo was the venue for three oratorios. Colonna returned to Modena with *L'Assalone* [13] and *Giudith* [14], Old Testament tales of love and warfare. In both, the clamour of battle is suggested by war-like arias, rather than by choruses. The third oratorio was Stradella's *S. Edita, vergine e monaca, regina d'Inghilterra* [16].

After Stradella's assassination in Genoa in 1682, Francesco II bought from the composer's nephew and other sources a large collection of his compositions, including scores of five oratorios. Of these, *S. Edita* was the first selected for performance in Modena. As the oratorio celebrates, in the form of a morality play, the faithfulness of an English queen, it touches upon the situation facing the duke's sister in London. The Modenese, as we have seen in Chapter 1, were well aware of the dangers facing Maria Beatrice, and one can see the choice of *S. Edita* both as a symptom of their concern, and as a means of relieving their own anxiety by contemplating the stoicism of a former English queen. In the oratorio Edith (S.) is befriended by Humility (S.). After withstanding the temptations dangled before her by Pleasure (B.), Greatness (A.), Beauty (T.), and Nobility (S.), she returns to embrace Humility, who utters the words of the Royal Prophet (King David the psalmist) in a fitting conclusion, 'Who sows in tears, reaps in joy.' The only surviving score of *S. Edita* in the Biblioteca Estense is a duplicate copy made for the singers from the full score (now lost). It preserves the vocal parts, but not the instrumental ritornellos. In 1684, of course, Vitali had access to the full score, the duplicate score, and the parts; they were all listed in the catalogue of the ducal library.

The second oratorio season of the interregnum, 1685, proved even more enterprising than the first. It featured two oratorios from Queen Christina's circle in Rome: Pasquini's *S. Agnese* [19] and Scarlatti's *S. Teodosia* [22]; two from Bologna: Perti's *Abramo vincitor* [20] and *Mosè conduttore del popolo ebreo* [21] (Giardini's *Vita di Mosè*, 4); one from Ferrara: Palermino's *Il trionfo della morte* [18]; and one from Modena: Gianotti's *La Maddalena pentita* [17].

Duke Francesco was an avid collector of musical scores. A thorough examination of his Stradella manuscripts (scrutiny of paper quality, watermarks, handwriting, etc.) in recent years has enabled scholars to track down where many of the scores originated. No doubt the process will continue and will be extended to include scores by other important composers like Pasquini and Scarlatti. We may, for example, be able to discover whether the duke's score of Pasquini's *S. Agnese* came directly from Rome where it was first performed in 1677, or by way of Ferrara where it was performed in 1679, or from the library of the Oratorians in Bologna where the work was listed in 1682,[9] or was a copy made in Modena for the 1685 season. Until the matter is settled by scientific investigation, the most sensible assumption is that the score came from Rome, from the composer himself. Pasquini may have sent it to Modena as a sample of his artistry to clinch the commission he completed in 1687, setting to music Giardini's *I fatti di Mosè nel deserto*.

Returning from speculation to matters of fact, it is certainly the case that Francesco II purchased more scores than he ever intended to see performed in Modena. He probably sampled them all in private, but less than a third of the operas he owned reached the public theatres of Modena or Reggio. Oratorios, much cheaper to produce, fared better; on current reckoning almost 80 per cent were given a public airing. To recognize that the duke was a bibliophile is crucial to our understanding of Modenese oratorio. From 1677 to 1684 every oratorio performed in Modena had had a local connection but, of the forty-five oratorios produced in the period 1685–9, almost half were composed by musicians who never set foot in Modena. Their works were collectors' items which the duke made available for performance is San Carlo rotondo. Thus, for five seasons, the Congregazione di San Carlo had the good fortune to hear oratorios from every thriving production centre in Italy.

Pasquini's *S. Agnese* had been composed in Rome in 1677 in collaboration with the aristocratic poet Benedetto Panfilio. The Panfilio family, one of the wealthiest in Rome, had commissioned Borromini to build the church of Sant' Agnese beside their palace, as a splendid

---

[9] See O. Mischiati, 'Per la storia dell 'oratorio a Bologna: tre inventari del 1620, 1622, e 1682', *Collectanea historiae musicae*, 3 (1963), 131–70.

centrepiece to the Piazza Navona. Work on the exterior of the church had ended with the raising of the twin towers in 1666, but the interior decorations were not to be finished until the end of the century. In its Roman context, therefore, the oratorio had a special significance for the Panfilio clan: it commemorated a Roman saint in whose honour they had raised one of the finest High Baroque churches in Europe.

The subject-matter of the oratorio, the martyrdom of a virgin, was extremely popular in Rome. With the support of Queen Christina, and to the delight of vain castratos who landed the title-roles, innocent victims of pagan oppression from the days of imperial Rome were fondly remembered in the oratories. Agnes, Cecilia, Teodosia, and many others were honoured for their chastity, and for their fortitude in the face of torture and death.

Panfilio's libretto traces the legend of the saint: her refusal to follow her mother's wishes and marry Flavio, her decision to dedicate her virginity to God, her desire for martyrdom, and her execution by a stabbing in the throat. At the end, her mother and Flavio are left to reflect on her fate; 'There you lie, Agnes, free at last from the fear of death.' The text comprises recitatives, arias in a variety of moods, and four duets. Pasquini's music, with its warm harmony, shapely phrasing, and mellifluous counterpoint seldom fails to please the ear. Conservative by nature and an avowed disciple of Palestrina, Pasquini persists in using the strophic aria form and employs fioritura with much greater restraint than his young contemporary Scarlatti. The lack of ostentation in the vocal writing is matched in the instrumental music, scored for two violins, violetta (with mezzo-soprano clef), viola, and basso continuo. Only one aria and one duet are accompanied throughout by this ensemble; the rest of the formal numbers are tastefully decorated with short ritornellos and interludes for the instruments. Pasquini's genial personality, expressed in music of great charm, found an appreciative audience in Modena. Together with Scarlatti, whose *S. Teodosia*[10] was also heard in the 1685 season, he brought a new aesthetic, a style of subtle shades of emotion, into the Lenten devotions of the Modenese.

Giacomo Antonio Perti of Bologna was only 24 when he made his debut at San Carlo rotondo with the fourth episode of the Moses cycle, *Mosè conduttore*, and a work he had composed in December 1683 for the oratory of Santi Sebastiano e Rocco in Bologna, *Abramo vincitor de' proprii affetti*.[11]

---

[10] For details of the oratorio, see Ch. 7.

[11] The score of *Abramo* at Modena is entitled *Agar scacciata*. The oratorio was renamed when it was presented again in Bologna in 1689 in honour of Prince Ottoboni, the nephew of Pope Alexander VIII. A score with the title *Abramo vincitor* survives in the Archive of San Petronio, Bologna, cat. no. P. 55. 1.

The libretto of *Abramo* explores the relationship between Abram (B.) and his two wives Sara (S.) and Agar (S.). The plot calls for a good deal of ensemble writing. At the end of Part I Abram and Agar share a touching love-scene in which a brief duet, 'Per opra d'amor/ Penoso contento', recurs as a refrain. The misery of the rejected wife Agar is expressed in an aria accompanied by two solo violins, after which she is joined by an Angel (A.) and Abram for a trio, 'Che pensi?/Che cerchi?/Non so.' The final pages of the score contain a reconciliation duet for Sara and Abram and a lively chorus in which three voices are accompanied by instruments *a 4*.

The libretto of *Mosè conduttore*, unusually for Modena, contains a list of the singers who performed the oratorio. Three came from the court at Mantua: Ferdinando Chiaravalli (A.)—Moses, Giovanni Bozzoleni (T.)—Hebrew Captain, and Francesco Ballerini (A.)— Narrator. The fourth was a member of the Modenese cappella, Antonio Balugani (B.)—Pharaoh, and the fifth, Domenico Cecchi (S.)—Egyptian General, is described as coming from Cortona. The list shows that the duke engaged guest singers for the oratorio season in Modena, much as theatre managers did when arranging seasons of opera. Perti's instrumental requirements for the oratorio are strings *a 4* and a solo trumpeter. The D trumpet is used in the opening sinfonia and in Pharaoh's aria 'Da fiere megere' in Part I. The display of instrumental and vocal virtuosity in Perti's setting compensates for the lack of visual spectacle in the presentation of this most spectacular episode, the crossing of the Red Sea.

Little is known about Padre Palermino (Bonaventura Aleotti), a Sicilian composer who spent the middle years of his career in Ferrara. One glance at the scores of his oratorios, however, reveals work of outstanding quality, for both dramatic flair and technical skill. The librettos he chose to set to music all contain allegorical characters. Though they arrest the flow of the drama, these characters are so colourfully portrayed that one is prone to welcome their intrusion.

Palermino's first Modenese oratorio, *Il trionfo della morte*, tells the story of the expulsion of Adam and Eve from the Garden of Eden. It opens with a passionate love-scene between Adam (T.) and Eve (S.), which is interrupted by Reason (S.) and a three-part Chorus of Virtues (S. S. A.). After giving Adam some sound advice, they leave the stage to Death (A.), Passion (B.), Lucifer (B.), and a Chorus of Demons (S. A. T. B.) who cajole and frighten Eve into submission. God (B.) appears in part II with a Chorus of Angels (S. S. A. T. B.) to call the couple to repentence, before expelling them from Paradise. The librettist describes the piece as an oratorio for five voices, 'con due Bassi di drammatica occurenza', perhaps indicating that Lucifer and God are to make spectacular entrances, like the deus ex machina familiar on Venetian stages, even if the oratory lacked the stage machinery to do them justice.

Palermino's masterly handling of five-part string textures and choral fugues in *Il trionfo della morte* stands in stark contrast to the mundane workmanship of Antonio Gianotti, whose *La Maddalena pentita* was presented on the Tuesday of Holy Week, 1685. Gianotti, who joined the court orchestra in 1687, had only an elementary knowledge of composition: da capo arias with motto beginnings are trotted out with monotonous regularity. Perhaps the fact that the composer was titled (Magnifico) persuaded the duke to give this drab piece an airing.

On 23 April 1685 the duke's sister was crowned queen of England. Having suffered eleven years of hostility and vilification in England, Maria Beatrice was now safely enthroned. Jubilation was the order of the day in Modena. Giardini published an ode and eight sonnets to mark the occasion. Vitali composed a cantata for a special meeting of the Accademia de' Dissonanti. But for Duke Francesco and his subjects there were two clouds on the horizon, fast approaching. The duke of Monmouth (Charles II's illegitimate son) invaded England with a Protestant task force to unseat James II, a rebellion which came to grief on the battlefield of Sedgemoor. The Abbé Rizzini wrote to Francesco on 27 July describing the débâcle and Monmouth's vain pleas to the queen for clemency:

he [Monmouth] wrote the same day, Tuesday, to the Reigning Queen to plead for his own life; but Her Majesty replied to him that if she alone had been offended she would have pardoned him gladly; but as it was a question of offences against the King and the State, she could not, and would not, interfere at all.[12]

Monmouth was summarily executed, and that cloud dispersed. The other threat was from the wrath of Louis XIV. The French king had been deeply offended by Prince Cesare Ignazio's effrontery in marrying his own sister, Angela Maria Caterina d'Este, to Emanuele Filiberto of Savoy in defiance of French wishes. Louis had enlisted, and received, the support of the queen of England in this affair; she had no time for her upstart cousin. The upshot of it was that Cesare Ignazio was forced to leave Modena on 23 June 1685, and spend a year in exile at Faenza. Francesco was devastated, and complained bitterly of fatigue and overwork until his cousin returned to court in the autumn of 1686.

The oratorio season of 1686 was full of delights: three works from

[12] See C. de Cavelli, *Les Derniers Stuarts à Saint-Germain-en-Laye* (Paris, 1871), ii, doc. 71.

Bologna, two from composers employed at the Mantuan court, one from Rome and, finally, one composed by Vitali himself, his first since leaving Bologna.

What drew Vitali back to oratorio composition was the chance to write one about Monmouth's rebellion and defeat in England. The score of *L'ambitione debellata overo La caduta di Monmouth* [28] names the five participants in the drama as Reason (S), Faith (S.), Ambition (B.), Treachery (T.), and Innocence (A.). In the libretto, however, they are identified as Narrator, Queen of England, Monmouth, Duke of Argyll, and King of England respectively. The oratorio deals only with the personal conflicts that arose during the rebellion. Part I is a battle royal between the Queen and Monmouth, she a fearless defender of the faith, he an aggressive and dangerous upstart. Their quarrel culminates in a musical duell: heated exchanges in recitative lead to an aria of two stanzas in which the fiery *passaggi* of Monmouth's 'Il mio brando fulminante sul tuo capo piomberà' provoke an equally brilliant response in the Queen's 'E l'audacia tua spietata calamità di saette'. An agitated duet over a running bass brings Part I to a close. The rest of the oratorio traces Monmouth's downfall. After the news of Argyll's failure to win Scotland, he resolves to fight on alone in an accompanied aria, 'Se mi manca il Tradimento', which Vitali turns into a sprightly gigue. His self-confidence is short-lived, however, and in the final pages of the score he adopts the demeanour of a penitent, while King and Queen sing of triumph in a suave duet. A well-crafted fugal chorus *a 5* ends the work.

Colonna revisited Modena in March 1686 to direct two of his finest oratorios, *Mosè legato di Dio* [25] (*Vita di Mosè*, 3)[13] and *La profezia d'Eliseo* [24].[14] At this time Colonna was acting as a consultant to the duke on the upkeep of the royal collection of musical instruments. A passport issued by the duke on 9 April[15] curiously describes the holder Colonna as 'my maestro di capella'. It authorizes him to travel to Venice to purchase harpsichords for the court. Colonna's letters of advice on technical matters increased in number after the new maestro Gianettini arrived in May, a sure sign that his services were greatly appreciated.

Also from Bologna came a Passion oratorio by Pietro degli Antonii, *L'innocenza depressa* [23]. The Virgin Mary (S.), St John (A.), Mary Magdalene (A.), Nicodemus (T.), and Joseph of Arimathea (B.)

---

[13] For an analysis of the oratorio see H. E. Smither, *A History of the Oratorio*, i (Chapel Hill, NC, 1977), 327–32, and Crowther, 'Development of the Oratorio', 317–23.

[14] For details of the oratorio, see Ch. 8.

[15] See the Colonna file in *I-MOs* MM, Busta 1A.

gather at the tomb of Christ to console each other. Part I is lacrimose throughout, reaching a climax in Mary Magdalene's plaintive aria on a chromatic ground bass, 'Lasciatemi morire'. In Part II, grief turns to anger and bitterness until an Angel (S.) intervenes to announce that torment will give way to happiness. To the sound of the trumpet, the Angel sings his exit aria, 'Fra gioie esulti ogni cor', leaving the Virgin the final word in a joyful accompanied aria, 'Sù, sù, risvegliatemi'. Though degli Antonii was more famous as a virtuoso cornettist than as a composer, the score of *L'innocenza depressa* shows that he was a master of instrumental idioms, and a subtle colourist.

The one Roman oratorio heard in Modena in 1686 was Scarlatti's *Il trionfo della grazia overo La conversionce de Maddalena* [27], first performed in Rome in 1684. It is an allegorical work in which Youth (S.) and Penitence (A.) vie for the soul of Mary Magdalene (S.). Predictably, Penitence wins the contest, and in more ways than one. In addition to stealing the last word (the moral in recitative: 'Human desire may stray, but after a thousand slips, can find no peace unless it returns to God'), he also has the honour of singing the last aria and the last accompanied aria in the oratorio. Poor Mary! Even at the point of her conversion she is upstaged by the orchestra, which plays an extraordinary chromatic sinfonia to denote the movement of the heavens. Static in conception, *La Maddalena* is a charming if somewhat protracted oratorio, close in form and spirit to the chamber cantatas which flowed profusely from Scarlatti's pen throughout his long career.

At the time of writing *L'innocenza di Davide* [26], Carlo Ambrogio Lonati was on the payroll of the duke of Mantua. He visited Modena in 1686 to direct his own opera, *I due germani rivali*, and perhaps also that of his friend Stradella, *Il trespolo tutore*. As the librettist of his oratorio was Modenese, and the score shows signs of hasty workmanship, Lonati might well have dashed it off whilst on leave in Modena. The oratorio is based on the story of David's expulsion from the court of King Saul. The tone of the work is set by the opening Chorus of Hebrew Women, dancing as they sing, 'Di timpano e cetra con lieti concenti'. They reappear in festive vein on three later occasions, thus providing a choral framework for the more dramatic scenes in which Saul is driven to jealousy by an evil spirit, David is pursued by an angry mob of servants, and Jonathan and David make their fond farewells. This last scene must have been a painful reminder to the duke of his cousin, exiled in Faenza, from whom he had parted with great reluctance.

Marc Antonio Ziani was also in service at Mantua when his first oratorio *Giuditta* [29] was heard in Modena. In the same year, 1686, his opera *Tullo Ostilio* was produced at Reggio. Ziani's career as a composer had begun in the opera-houses of his native Venice in the mid-1670s where one of his contemporaries was a bass singer and organist in St Mark's, Antonio Gianettini. As Ziani's oratorio was being sung in San Carlo rotondo, Gianettini was preparing to leave Venice to take up the vacant post of maestro di cappella at the Modenese court and shoulder some of the responsibilities that Vitali had borne for three years with notable success.

6

De Grandis's *Il nascimento di Mosè* (1682)

Il nascimento di Mosè, an oratorio in two parts
Music by Don Vincenzo de Grandis
Libretto by Giovanni Battista Giardini
Characters: Testo (Narrator)   A.
            Faraone (Pharaoh)   B.
            Generale (General)   Bar.
            Consigliero (Counsellor) S.
            Madre (Mother of Moses) S.
            Coro *a 4* (Chorus)   S. A. T. B. or S. S. T. B.
Instrumental accompaniment *a 3*
First performed in the oratory of San Carlo rotondo, Modena, in May 1682.

The Libretto

Although *Il nascimento di Mosè* was the first episode in *La vita di Mosè*, the poet makes no mention of his grand cyclical scheme in the printed libretto of 1682, nor in the libretto of the second episode, *Il matrimonio di Mosè*, of 1684. One can only surmise that the omission was tactical, the author testing the court's reaction to his project before declaring his intentions in episode 3, *Mosè legato di Dio* (1686).

Dedicating his work to the duke, Giardini writes, 'I place at the feet of Your Serene Highness the baby Moses, who is no sooner born than exposed in such a fragile craft to the fury of the waves and the power of the winds'. With a flattering allusion to the duke he continues, 'He has such great need of a Pole Star that will guide him through the storm and protect him from the threat of shipwreck.'

The primary source of the oratorio is Exodus 1: 8–2: 10. The basic pattern of the biblical narrative is preserved in Giardini's dramatization. Pharaoh fears the expanding population of the Hebrews within his realm (1: 9). He imposes hard labour upon them (1: 13) and commands the Hebrew midwives to strangle all male children at birth (1: 16). When that fails, he decrees that all new-born sons are to be thrown into the river Nile (1: 22). Moses is born (2: 2). His mother hides him in the house (2: 2) and then places him in a wickerwork basket in the river (2: 3). He is discovered by Pharaoh's daughter (2: 5), who brings him up as her own son at court (2: 10).

But these events are only the bare bones of Giardini's drama. To enliven the story, he creates a Pharaoh who is tyrannical and paranoid, suffering from nightmares and hallucinations, a captain of the royal bodyguard (General) who is vengeful and bloodthirsty, and an Egyptian Counsellor who flatters his royal master and shows utter contempt for the Hebrews. The role of the Mother of Moses as an innocent victim of tyranny is highlighted in two impassioned soliloquies in Part II. Linking the scenes of cruelty and oppression is the Narrator, who introduces the characters, describes the events, and offers (together with the Chorus) a moral commentary upon them. Like the Narrator in *Susanna*, he speaks for Giardini.

Entertainment and instruction are the poet's objectives. He achieves the first by using a vocabulary rich in extended metaphors, and the second by delivering a strong religio-political message, that tyranny and intolerance (antisemitism) can only lead to disaster in a world governed by God's righteous will. To ensure this latter objective, he allows only the moralizing voices of the Narrator and the Chorus to be heard in the final section of Part II.

The dramatic structure of the oratorio is as follows:

Sinfonia
Part I:

Narration 1	Narrator: Rulers are paralysed by internal fears. Pharaoh is perplexed by the growing strength of the Israelites.
Scene 1	Pharaoh: Alone in his bedchamber, Pharaoh describes his nightmares in a long soliloquy.
Narration 2	Narrator: The General and the Counsellor rush in to comfort their demented sovereign.
Scene 2	General, Pharaoh, Counsellor, Chorus: The General vows to kill Pharaoh's tormentors. Pharaoh has a vision of a baby who will overthrow him. He is frightened even by the colours (gold and purple) of his own regalia. The Counsellor and General advise, and report on the effect of, repressive measures against the Hebrews. Pharaoh orders a public proclamation, that all new born Hebrew males are to be cast into the Nile. The Chorus warns harsh politicians not to found their empires on falsehoods.

Part II:

Narration 3	Narrator: Pharaoh's decree strikes panic among the

Hebrews. The Mother of the new-born Moses is introduced.

Scene 3	Mother: She laments over the fate of her baby.
Narration 4	Narrator: A commentary on Pharaoh's obduracy.
Scene 4	Pharaoh, Counsellor, General: Pharaoh, his spirits revived by the blood-letting, discusses with his advisors the necessity of pruning to maintain the health of the realm.
Narration 5	Narrator: He describes Moses' embarkation.
Scene 5	Mother: She bids farewell to her son, committing him to God's protection.
Scene 6	Narrator, Chorus: The Narrator tells of Moses' rescue by the daughter of Pharaoh. A vengeful Chorus questions her wisdom, and later her father's, in welcoming the baby to the Egyptian court. The Narrator's final comment is that kings who fancy they control events prove to be mortal, whereas God's rule is eternal.

As an oratorio librettist, Giardini uses figures of speech much more flamboyantly than his contemporaries. In Scene 1, for example, Pharaoh begins to describe his nightmares in the kind of terms other poets had used to depict the temptations of St Antony:

> Dai chiostri della Morte
> Erinni scatenate.
> Demoni, fuor usciti.
> Che di larve sognate la mia mente
> Impremete, trovarte, se sapete,
> Nove forme di pena.
> E in duol'eterno
> Accendetemi in petto un vivo Inferno.

(From the cloisters of death, break out, ye Furies! Come out, ye demons! You spectres who dwell in my mind, devise, search out, if you know of them, new forms of punishment. In my breast, set alight a living inferno of endless pain.)

But as the soliloquy progresses, conventional phraseology gives way to more paranoid ravings:

> Come si fosse il Talamo regale
> Nidata di Ceraste
> Se non pungoli, e tosco
> Io non vi trovo.
> A torturar la membra
> Paion ruide Furii

Da fuso Etiopeo filati bissi.

(The royal marriage-bed seems like a nest of vipers, stinging and poisoning in secret. To torture my limbs it seems that rough Furies are spinning me around by a black shank.)

No less remarkable is the infusion of metaphors drawn from horticulture in Scene 4:

COUNSELLOR: Piange afflitto Israelle
                     E, da provida scure,
                     Ogni suo germe tronco in erba ravvisa.

(Afflicted Israel weeps, and all its immature seed has been identified and cut off by the provident axe.)

GENERAL: Sì, sì, ma più rigoglio
                     Si mette i palmiti suoi vite recisa.

(Yes, yes, but vines grow more luxuriant when their shoots are pruned.)

COUNSELLOR: Dal duro giogo oppressa,
                     Sorger non puo più la fortuna Ebrea.

(Oppressed by a heavy yoke, Hebrew fortunes cannot rise again.)

GENERAL: Al hor che più s'aggrava,
                     Sorge con più vigor Palma Idumea.

(When the Idumean palm is overburdened it grows more vigorously.)

COUNSELLOR: Cadde, squarciato il seno
                     Dela sette nemica,
L'ardir sognato e la superbia insana.

(The torn breast of the enemy sect is fallen, its boldness a dream, its pride insanity.)

PHARAOH: Dal suo lacero sen, la mela grana
                     Prende maggior vaghezza.
                     E solo al hora mostra
                     Nel fianco aperto del vegeto rubino
                     La Natia grana e l'ostro suo più fino.

(The pomegranate takes great beauty from its lacerated breast. Only then does it show in its cloven side the natural grains of luscious ruby and finest purple.)

COUNSELLOR: Che più dunque si bada?
                     Su le teste rubelle
                     Scenda il colpo fatal dela tua spada.

(Then why delay? Let the fatal stroke of your sword fall on the heads of the rebels.)

If likening the heads of Hebrew babies to pomegranates seems callous, the remarks, taken in their dramatic context, highlight the author's view that antisemitism is a cruel sport. Infanticide, to those who perpetrate it, may appear no worse than a judicious pruning of the garden. The effects of such cruelty, however, are starkly shown in the agonized plaints of the Mother in Scene 3:

MOTHER: Che intesi? Ahime! Che intesi?
                          E sara vero, ò Dio,
                          Che di mia speme il fior
                          Secchi si tosto?
                          E che il nato bambino,
                          Con efimera vita
                          Uniforme à la rose,
                          Habbia il destino?
                          Perchè nacesti mai,
                          Pegno di questo cor,
                          Se poi dovevi cedere
                          À i colpi del rigor?

(What did I hear? Alas! What did I hear? And will it be true, O God, that the flower of my hope must wither so soon? And that my new-born baby will have an ephemeral life, like the rose, as its fate? Why were you ever born, Pledge of my heart, If then you had to yield To the blows of rigour?)

Scenes like this move us to pity, but offer no solution to the problem of tyranny. For that we have to wait until the final scene of the oratorio. As Pharaoh welcomes the infant Moses into his household, the Chorus challenges him and prophesies:

                          Che fai, misero, che fai?
                          Prendi à latto quella Tigre
                          Che paventi.
                          Da cui denti
                          Lacerato,
                          Divorato,
                          Un dì sarai.
                          Che fai, misero, che fai?

(What are you doing, wretched man, what are you doing? You are drawing to your side that tiger which you feared. One day you will be mauled and devoured with his teeth. What are you doing, wretched man, what are you doing?)

In the same mood, but with a rapid change of metaphor, the Narrator adds:

                          In lui visse quel sole
                          A di cui raggi seccò l'Egizia speme,
                          E rifiori la libertà Giudea.

(In him [the young Moses] lived that sun whose rays dried up the hopes of Egypt, and made Jewish liberty flower again.)

By expressing such high hopes for the young Moses, Giardini was, perhaps, preparing the ground at court for a sequel.

The Score

The oratorio was designed for performance by five solo voices. The four characters involved in Part I sing the concluding chorus. When the second soprano (Mother) joins the cast in Part II, the Chorus is reconstituted to include the Mother and release the alto for duties as Narrator in the final scene. A curious feature of de Grandis's scoring is that the music for the General, a baritone, is written at times with a tenor clef (his first recitative) and at others with a bass clef (his first aria).

The instrumental requirements of the oratorio are two violins (changing to violettas for the Mother's accompanied recitative in Part II) and basso continuo. The Sinfonia is a trio sonata in three sections: a solemn opening in 4/4, a triple-time dance in binary form, and a lively canzona on a bustling subject reminiscent of Frescobaldi (Ex. 6.1). Though tonally unadventurous, the Sinfonia is of ample dimensions: 131 bars.

Ex. 6.1

There are 17 arias in the oratorio of which 10 are accompanied by the violins. Giardini's lyrics explore a variety of moods, but their poetic form precludes much use of the increasingly popular da capo aria form. Consequently, only 4 arias are in da capo form; the rest are through-composed (12) or strophic (1). Three duets, all in bellicose mood, and two quartets (choruses), the second of which is repeated with adapted lyrics, complete the array of formal numbers. These are distributed among the five soloists as shown in Table 6.1.

The letters of de Grandis preserved in the State Archive reveal that he was passionate and impulsive by nature. Thus, as a composer of

TABLE 6.1. Distribution of Numbers in *Il nascimento di Mosè*

	Counsellor S. 1	Mother S. 2	Narrator A.	General Bar.	Pharaoh B.
Continuo arias (7)	0	2	1	1	3
Accompanied arias (10)	2	2	3	2	1
Duets (3)	3	0	0	2	1
Quartets (3)	3	2	1	3	3

oratorios, he responded best to highly charged dramatic situations which called for expressions of fury, cruelty or heartbreak. In the General's first aria, 'Chi è l'autor della tua pena', the violins express the soldier's fury while he, in measured and courteous terms, suggests to his sovereign a rigorous military solution to his problems (Ex. 6.2).

Ex. 6.2

Ex. 6.2 (*cont.*)

(He who is the author of your distress . . .)

The introduction to this aria begins, somewhat eccentrically, on the supertonic of B minor, in order to run smoothly from the foregoing recitative which ends on a chord of F sharp. Such linking devices between formal sections abound in de Grandis's setting.

The duet 'Non si desista', sung by the Counsellor and Pharaoh, occurs at the climax of Scene 4 (Ex. 6.3). Their declaration of a vendetta against the Jews is sustained for fifty-nine bars. Following on its heels is a scene of contrasting emotion: the Mother bids farewell to her infant son. Her soliloquy reaches a point of great anguish in the recitative 'Mà, ohimè, ch'io t'abbandoni', accompanied by two

Ex. 6.3

Ex. 6.3 *(cont.)*

nò, nò, nò, nò, nò, da - la ven - det - - - -

si de - sis - ta nò, nò, nò, nò nò da - la ven-det - -

- ta non si de - sis - ta non si de - sis - ta nò

- - - - - ta non si de - sis - ta nò

(One should not desist, no, no, from taking revenge)

violettas (Ex. 6.4). Her cries of despair, expressed through chromaticism and discord, dissolve into a grief-laden arioso, characterized by simple imitative patterns arranged in falling sequences.

Ex. 6.4

violettas

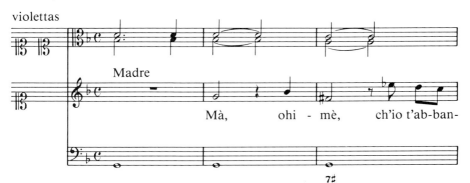

Madre

Mà, ohi - mè, ch'io t'ab-ban-

Ex. 6.4 (*cont.*)

Ex. 6.4 *(cont.)*

ni-ma mi - - a.

6   5   4 3

(But, alas! if I were to abandon you, if I were to leave you without dying . . .
too harsh is the pain of it, so much of it promises to overwhelm my soul
with grief.)

    The oratorio ends with a recitative and continuo aria for the
Narrator. The musical setting of the aria seems strangely at odds with
the text. Giardini's lyric sternly reminds tyrannical princes of their
mortality, perhaps calling for an aria di bravura. De Grandis responds
to it, however, by writing an aria in the style of an elegant sarabande
in G major, replete with languishing hemiola cadences. He changes
metre and mood (4/4, 'grand e adagio') to point up the threat of royal
extinction in the last line of the poem, but the general charm of the
music is disconcertingly inappropriate (Ex. 6.5). The reason for the
mismatch may lie in the fact that, unlike Giardini, de Grandis was a
newcomer at the Modenese court. Perhaps he felt uneasy at ending his
first large-scale composition for the duke on a harsh note that might

Ex. 6.5

Testo

Or voi   che vi fon - da - te, che

bc

Ex. 6.5 *(cont.)*

(Now you who base yourselves on erroneous principles . . .)

have caused offence. By softening the impact of Giardini's strictures, he secured a lieto fine for himself.

During their working relationship at court, Giardini and de Grandis formed a close friendship. After leaving Modena, de Grandis sent a stream of letters (thirty-one are extant) to Giardini between June 1687 and May 1694, which show the composer's continuing interest in Modenese affairs.

7

Scarlatti's *S. Teodosia* (1685)

S. Teodosia, an oratorio in two parts
Music by Alessandro Scarlatti
Libretto—anon.
Characters: Teodosia (martyr)   S.
             Decio (royal counsellor)   A.
             Arsenio (prince)   T.
             Urbano (emperor)   B.
             Chorus *a 4*   S. A. T. B.
Instrumental accompaniment *a 4*
First performed in Modena in 1685

The Sicilian composer Alessandro Scarlatti (b. Palermo, 1660; d. Naples, 1725) first achieved fame in Rome under the patronage of Queen Christina and her circle, following the success of his opera, *Gli equivoci sembiante*, in 1679. His close association with Bernardo Pasquini and the Panfilio family may account for his links with Modena. Under Vitali's directorship, *S. Teodosia* (anon., Scarlatti), *S. Agnese* (Benedetto Panfilio, Pasquini), and *Il trionfo della grazia* (Panfilio, Scarlatti) were presented in Modena in the seasons of 1685 and 1686. *S. Teodosia* was Scarlatti's sixth oratorio and, as far as we know, the first of his to be premièred outside Rome.

The Libretto

Though the libretto of *S. Teodosia* was printed by the ducal press of Soliani, it does not appear in the *Raccolta d'oratorii* associated with the oratory of San Carlo rotondo. Thus, we cannot be sure of the location of the performance in Modena. The libretto lacks a dedicatory address, so the circumstances of the performance remain a mystery.

The anonymous librettist of the oratorio was a Roman through and through. His subject-matter, extracted from the popular hagiographies of the day, is the martyrdom of a Christian virgin in pagan Rome. Teodosia's refusal to marry Prince Arsenio, son of the tyrannical Emperor Urbano, leads her to protracted torture and, finally, death by the sword.

In handling this gruesome legend, the librettist fulfils all the conditions prescribed in Arcangelo Spagna's famous treatise *Il discorso dogmatico* (Rome, 1706), for creating a 'perfect spiritual melodrama'. He abandons the role of Narrator, allowing the story to unfold through the words of the four protagonists. He observes the Aristotelian unities of time, place, and action. The successive phases of Teodosia's temptation, torture, and death correspond to the protasis, epitasis, and catastrophe of classical tragedy. With regard to prosody, the rhyming texts of his arias are short, and cast in a form suitable for da capo treatment. His recitatives also make effective use of rhyme. In the first speech of Decio (addressed to the lovelorn Prince Arsenio), we can observe a combination of blank verse, short rhyming couplet, and heroic couplet:

> Lascia, Signor. Al mio pensier l'incarco,
> Che di piegar alle tue giuste voglie,
> Di Teodosia l'affetto
> Io ti prometto.
> E lieto più sù la mia fè riposa
> Che Teodosia vedrai hoggi tua sposa.

(Stay, my Lord. Trust in my counsel. I promise that Teodosia's affections will submit to your just desires. You can happily rely on me that today you will see Teodosia your betrothed.)

Italian recitatives were normally cast in lines of eleven or seven syllables but here the librettist artfully shortens the fourth line to five syllables to emphasize Decio's personal pledge. Stylistic analysis of oratorio texts may eventually reveal the identity of the author of *S. Teodosia*; for the present, however, we must be content with the knowledge that he belonged to the progressive, pro-Spagna, group of Roman poets.

Having dispensed with the role of a narrator, the librettist presents the oratorio in five dramatic scenes. The first three comprise the protasis, Scene 4 the epitasis, and Scene 5 the catastrophe:

Sinfonia
Part I:
Scene 1　Arsenio, Urbano, Decio: Prince Arsenio, grief-stricken by his unrequited love for Teodosia, is comforted by his father and the counsellor. Decio promises to persuade the maiden to submit before the day is out.

Scene 2　Decio, Teodosia: Decio's embassy is scorned by Teodosia, whose heart is full of heavenly love. She will not submit.

Scene 3　Teodosia, Decio, Arsenio, Urbano: Teodosia is interrogated by her three adversaries: the ever-tearful suitor, the bombastic emperor threatening dire penalties for disobedience, and the wily counsellor trying to placate the prince and tempt the saint. At the end of the exchanges Teodosia declares, 'I wish for death.'

Part II:

Scene 4   Urbano, Teodosia, Arsenio, Decio: The emperor calls in his ministers of justice to begin the trial by torture. Teodosia welcomes their ministrations with open arms. Arsenio is sickened by the spectacle, Decio amazed to see her so happy under duress. The saint fixes her thoughts on heavenly bliss.

Scene 5   Teodosia, Chorus: With her dying breath, Teodosia challenges her royal persecutors, and commends her faithful soul into the hands of the blessed spirits and the angels of heaven. The Chorus points the moral with a spiritual epigram: 'To the one who dies for God, death is life.'

The drama unfolds in a sequence of alternating recitatives and arias, broken only by the occasional ensemble. In Scene 2, for example, Teodosia politely rejects Decio's pleas in a decorous duet:

DECIO: Bella, perchè disprezzi
                Chi t'offerte in dono il cor?
TEODOSIA: Lo sprezzo sol che pago
                D'amor celeste è il sen.
DECIO: Perchè non cedi ai vezzi
                Con che t'alletta amor?
TEODOSIA: Amor del tuo più vago
                Soggetto il cor mi tien.
DECIO: E con folle desio
                Tu sdegni un Prence.
TEODOSIA: E con fedel desio
                Per amare un Dio.

(Beautiful lady, why do you despise him who gives you his heart? ‖ I scorn him only because my breast is full of heavenly love. ‖ Why not yield to the charms with which love entices you? ‖ A love more attractive than yours enthralls my heart. ‖ With foolish longings you scorn a prince. ‖ With faithful longings to love God.)

As the catastrophe approaches, all pretence at civilized debate is thrown to the winds. Teodosia, *in extremis*, is locked in mortal combat with Arsenio and Urbano in an agitated trio:

TEODOSIA: Costanza ci vuol per combattere

ARSENIO/URBANO: Fierezza ci vuol per combattere
TEODOSIA: Costanza chi brama abbattere
                La crudeltà.
(Constancy is needed to fight. ‖ Ferocity is needed to fight. ‖ Constancy is needed to beat down cruelty.)

After such bellicosity, the simplicity of Teodosia's farewell to the world, and the measured pronouncement of the final Chorus are breathtaking:

TEODOSIA: Angeliche schiere
             Prendete il mio cor.
             A voi già ne vola,
             Cingetelo, ò stelle,
             Più bella d'ardor.
             Angeliche schiere
             Prendete il mio cor.
CHORUS: Di Teodosia il martir
             Chiaro t'addita
             À chi more per Dio,
             La morte è vita.
(Angelic host, take my heart. Already flying to you, embrace it, O stars, in your radiant glow. Angelic host, take my heart. ‖ The martyrdom of Teodosia shows you clearly that, to the one who dies for God, death is life.)

The oratorio is, in essence, an apologia for chastity and Christian fidelity. Though some of the argumentation in Scene 3 is rather protracted, there is enough human interest in the drama, and quite enough realism in the trial by torture, to bring the play of stock characters alive, and to elicit real sympathy for the innocent victim at the centre of events.

The Score

The extant materials relating to Scarlatti's setting of S. *Teodosia* in the Biblioteca Estense are a full score (including instrumental parts) at Mus. F. 1058, a score containing only the four vocal parts at Mus. F. 1059, and individual part-books for Arsenio and Urbano.[1] The full score shows that the string players are involved in all the formal numbers, excepting a brief quartet in Part I.

The Sinfonia functions as a brief curtain-raiser to Arsenio's accompanied aria 'Se il mio dolore'. There are no thematic connections

[1] See A. Chiarelli, *I codici di musica della Raccolta Estense* (Florence, 1987), items 666, 26, 279, 439.

between the two pieces, despite the fact that they share the same key and metre, and their textures are similar.

'Se il mio dolore' is the first of 17 arias in the oratorio. Ten are in ternary form, 4 are through-composed, and 3 are too brief to merit formal classification. An interesting feature of the libretto is that it calls for 10 arias with two matching stanzas. Scarlatti handles such lyrics economically by utilizing the same music for both stanzas. His strategy is quite acceptable when the basic emotions expressed are similar, as in Teodosia's 'Mi piace il morire/E dolce il tormento', but when Arsenio is required to sing the contrasting affections of 'Oh lieto quel core/E sempre dolente' to the same music, one is bound to question the composer's judgement. Perhaps like Dowland, whose stanzaic lutesongs raise similar problems for the interpreter, the young Sicilian left the problem to be resolved by performers of good taste. The formal numbers in the oratorio are distributed as shown in Table 7.1.

TABLE 7.1. Distribution of Numbers in *S. Teodosia*

	Teodosia S.	Decio A.	Arsenio T.	Urbano B.
Continuo arias with codas (6)	2.5[a]	1	1.5[a]	1
Accompanied arias (11)	6	1	2	2
Ensembles (4)	4	3	3	3

[a] Arsenio and Teodosia sing a stanza each of the aria 'Se vuoi crudel' in Part I.

The variety and quality of Teodosia's arias ensure that the castrato dominates the performance. In the middle of Scene 3, Teodosia is involved in a contest of wills (and of vocal virtuosity) with Urbano. He expresses his fury in the aria 'Già d'ira m'accendo' (Ex. 7.1), accompanied by the strings in stile concitato. She responds in the next aria, 'Son pronta all'offese', in like manner (Ex. 7.2). Her anger is expressed in messe di voce over a thundering bass, while the upper strings give spondaic support to her defiance. These settings are perfectly designed to show off the technical skill of bass and castrato voices respectively.

By contrast, Teodosia's prayer at the end of Scene 4, 'Soccorretemi, Cieli fedeli', is an anguished cry from a soul in torment. For this

Ex. 7.1

(Already I am inflamed with anger.)

Ex. 7.2

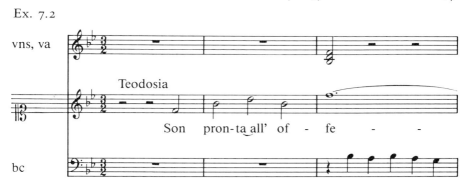

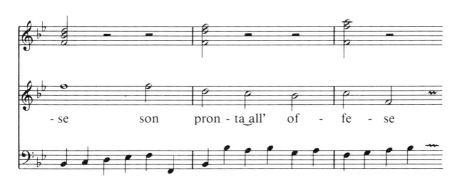

(I am ready for insults)

lament (Ex. 7.3), Scarlatti uses every device at his command to achieve emotional intensity: suspensions, phrase extensions, feminine cadences, telling silences, chromatic harmony, echo effects and, finally, a long messa di voce projected against rising scalic figures in the orchestra to depict her longing for death.

Scarlatti's handling of recitative is masterly. Typical of his artistry is a passage in Scene 4 where Urbano and Arsenio are discussing the fate

Ex. 7.3

(Succour me)

of Teodosia (Ex. 7.4). Urbano's advice to his son, to dry his eyes and cheer up, is subtly underlined by a change of key from D minor to F major. Arsenio's melancholy reply takes us to G minor, from where his father's kindly warning about the ferocity of beautiful tigers progresses to a cadence in B flat major. Up to this point, Scarlatti has supported the recitative by sustained semibreves and minims in the basso continuo. With Arsenio's next speech, however, the continuo

Ex. 7.4

Ex. 7.4 *(cont.)*

Ex. 7.4 (*cont.*)

(No longer ruin your face with grief. My son, leave the pitiless woman you love to die, and turn your gaze to a happier love. ‖ Ah, a dearer object for my thoughts than Teodosia I do not hope to find, father. ‖ The tiger is also beautiful but, by being so fierce, its beauty does not awaken love in the heart. ‖ It is true that the idol I worship is cold-hearted, but her cruelty still enamours me. ‖ Then what should be done to her? ‖ Let her live, live, not die.)

goes into regular motion (3/2) for the arioso 'Mà la sua crudeltà', and explodes with energy as the young lover offers his judgement on Teodosia's destiny: 'Viva, e non mora'. The tonal scheme gives 'viva' an F major colouring, 'mora' a G minor. The fact that 'mora' is negative (i.e. 'non mora') is an incidental inconsistency that goes unnoticed as Teodosia immediately sweeps into a D minor aria negating Arsenio's judgement: 'Mi piace il morire'.

The finest of Scarlatti's vocal ensembles frame the catastrophe. The trio, 'Costanza/Fierezza' (Ex. 7.5), is the most exciting number in the oratorio; a bravura display of three-part fugal counterpoint, with a thunderous orchestral coda. This angry exchange is a prelude to the saint's aria of leave-taking, which dies out in E minor. The final chorus, 'Di Teodosia il martir', accompanied by the strings, begins in a triumphant C major, affirming the glory of the martyr. At the words, 'A chi more per Dio' (Ex. 7.6), the tenor ushers in a solemn

Ex. 7.5

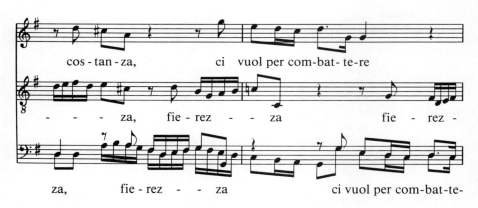

5

(Constancy is needed to fight. He who covets constancy . . . ‖ Ferocity is needed to fight.)

Ex. 7.6

(To the one who dies for God, death is life.)

fugue *a 4* in C minor, in which the countersubject, 'la morte è vita', is used with admirable consistency. The vocal polyphony is further enriched by the addition of three real parts for violins and viola, creating a web of seven-part counterpoint. The fugue comes to a close in the forty-fifth bar with a C minor chord on 'vita'. A lesser man than Scarlatti might have opted for a tierce de Picardie, but the master-craftsman of spiritual melodrama keeps us spellbound to the very end.

8

Colonna's *La profezia d'Eliseo* (1686)

La profezia d'Eliseo, an oratorio in two parts
Music by Giovanni Paolo Colonna
Libretto by Dottore Giovanni Battista Neri
Characters: Donna 1 di Samaria (Woman 1)   S.
              Donna 2 di Samaria (Woman 2)   S.
              Capitano di Joram (Joram's Captain)   S.
              Eliseo Profeta (Elisha the prophet)   A.
              Joram Rè di Samaria (King Joram)   B.
              Chorus *a 5*   S. S. A. T. B.
Instrumental accompaniment *a 8*
First performed in the oratory of San Carlo rotondo, Modena, in 1686.

Giovanni Paolo Colonna of Bologna (1637–95) was appointed maestro di cappella in the basilica of San Petronio, Bologna, in succession to Cazzati in 1674. Oratorio was familiar ground for him. He had served as organist under Carissimi in Rome in the 1650s, and held the post of maestro at the Oratorians' church, Madonna di Galliera, in Bologna from 1673 to 1688. *La profezia d'Eliseo* was his fifth oratorio to be presented in Modena.

Giovanni Battista Neri was a Bolognese author whose plays, pastorals, operas, and oratorios were performed in Venice, Bologna, Modena, and Reggio between 1677 and 1723. Among his musical collaborators were Albergati, Colonna, Gianettini, Giovanardi, and Vinacesi.

The Libretto

The source of Neri's libretto is the harrowing story of the siege of Samaria in 2 Kings 6: 24–7: 20. It is accurately paraphrased in a long *argomento* printed in the libretto and also, rather unusually, copied by hand into the score. In adapting the story for the oratory, Neri omits the episode of the lepers (discoverers of the deserted Syrian camp), but greatly expands the biblical account of the two women reduced to cannibalism by the severity of the famine in Samaria.

The dramatic structure of the oratorio is as follows:

Part I:
Prologue   Sinfonia *a 8* with duet for two sopranos: Samaria is besieged by the Syrian army. Famine and terror strike the city.
Scene 1   Woman 1, Woman 2: Two mothers make a pact to survive by devouring their own infant sons.
Scene 2   Joram, Captain, Elisha: King Joram and the Captain discuss their misfortune and accuse Elisha of not ameliorating God's anger. The prophet declares that the day will bring great plenty to Samaria. He predicts that the sceptical Captain will see it, but will not enjoy it.
Scene 3   Woman 1, Woman 2: Woman 1, filled with remorse, refuses to commit infanticide. She offers her own life to Woman 2 in place of her son's.
Part II:
Scene 4   Woman 1, Woman 2: Woman 1 resolves to save her son. She engages in another dispute with Woman 2.
Scene 5   Joram, Woman 1, Woman 2: The king arbitrates between them.
Scene 6   All: The denouement. The Captain announces that the seige is over. Woman 2 is exultant, Woman 1 and Elisha are vindicated. The Captain is trampled to death by the hungry crowd, thus fulfilling Elisha's prophecy. Joram's faith is restored.
Scene 7   Chorus: The Samaritans celebrate their new-found faith in the God of Israel.

Like the author of *S. Teodosia* (see Chapter 7), Neri had progressive ideas about the structure and style of oratorio librettos. He manages without a narrator, rhymes his recitatives, and observes the Aristotelian unities. After the general scene has been set in the prologue, the development of events is signalled indirectly by the dialogue. At the end of Scene 1, for example, Woman 2 says to Woman 1, 'But King Joram approaches, and with him the Captain. Elsewhere you may hear the terms of it [their pact].' As they retreat, Joram enters. Later, in Scene 5, however, for dramatic effect, the king's entry is unannounced and abrupt. He interrupts the women's squabbling by shouting 'Hey there, you base women! What's the reason for this strife?' From such details one can see that Neri, though writing for the Oratory, is as careful about such matters as pace and continuity as he would have been fulfilling an operatic commission; yet another sign of his progressive outlook.

All but two of Neri's aria lyrics incorporate a reprise of the opening lines, lending themselves to da capo treatment. His recitatives are more extended than was common in Rome (a symptom of Emilian conservatism), and much more lively. The cut and thrust of the mothers' dispute in Scene 4 sinks to the kind of banter one might overhear in a fish market. The life of Woman 1's son is at stake:

W. 1	S'a svenar la tua prole
	Tu nudristi nel seno
	Alma d'Inferno.
	Io non sono una Furia.
W. 2	Mendicato pensier.
W. 1	Saggio consiglio.
W. 2	Cedi il mal nato figlio.
W. 1	Ferma!
W. 2	Lascialo, indegna.
W. 1	Empia, che tenti?
	Sei folle.
W. 2	Vedrai. Apri quel petto.
W. 1	Taci!
W. 2	O ch'io lo svenerò, sì sì.
W. 1	Nò, nò, nol lascierò.
W. 2	L'ucciderò, sì, sì.
	Estinto in pasto
	Per sempre l'havrò.
W. 1	Vivente per sempre l'havrò.

(In butchering your child, you nursed in your bosom a hellish soul. I am no Fury. ‖ A beggarly though. ‖ Wise advice. ‖ Give up your ill-starred son. ‖ Stop! ‖ Let him go, worthless woman. ‖ Pitiless woman, what are you trying to do? You are crazy! ‖ You'll see. Open that chest. ‖ Shut up! ‖ Or I will slaughter him, yes, yes! ‖ No, no, I'll not give him up. ‖ I'll kill him, yes, yes! Dead in a meal for ever I'll have him. ‖ Living for ever I'll have him.)

This is not the sort of material one normally associates with the spiritual exercises of the Lenten season. Its lack of decorum, however, does indicate the lengths to which progressive librettists would go to rid themselves of pious clichés, and air their moral concerns in the language of the street. One should add that such realism in the oratory is a rare occurrence.

The Score

The maestro of San Petronio evidently relished setting *La profezia d'Eliseo* to music. For years he had tolerated the verbose librettos of

Bergamori. Now, with Neri's work, he had a professionally crafted text to stimulate his imagination and challenge his powers of invention. It brought the best out of him; *Eliseo* is undoubtedly his finest oratorio.

The scoring is for five solo voices and double orchestra. The range of voices (three sopranos, alto, and bass) is an unusual combination for its period, and poses a minor problem in the final chorus, as one of the sopranos has to sing the tenor part. Colonna cleverly masks the adjustment by keeping the tenor in a high enough register for him to continue using his unnatural voice without embarrassment. Professional finesse is also apparent in Colonna's handling of the orchestral music. For the patronal festivals in San Petronio, he was accustomed to writing music (e.g. concerted Masses) for large numbers of singers and instrumentalists, beside which the problems of scoring an oratorio for double orchestra must have appeared quite elementary. It did, however, involve more work than he gave to scoring for strings *a* 5 in his other oratorios, emphasizing once again Colonna's determination to do something special for *Eliseo*.

Colonna, knowing the resources at Modena, probably intended the parts to be played by stringed instruments (vn. 1, vn. 2, va., and b.c. × 2). The opening and closing numbers of the oratorio are in a style and key well suited to the participation of Bologna's natural D trumpets, but none of the treble-clef parts in the score fall within their restricted range of notes. Thus a double string orchestra offers the best realization. The two sections of the orchestra share the music on an equal footing. There is no hint here of the concertino/concerto grosso arrangement favoured by Stradella and other Roman composers. Though the textures in *Eliseo* have the mechanical drive of the Baroque concerto style, and in that respect are modern, the interplay of the two opposed groups is handled in a manner that would have been familiar to Carissimi or Virgilio Mazzocchi of an earlier Roman generation. There is virtuoso display in Colonna's instrumental writing but it is shared by all the instruments, not restricted to specialists.

The oratorio begins in an extraordinary way with what can best be described as a battle symphony for double orchestra in D major (Ex. 8.1); it carries no title in the score. Triadic motifs, dashing scales, and anapaestic rhythmic cells, rebounding from one group to the other, create a vivid impression of warfare. In bar 26, two sopranos join the fray, crying, 'Behold the enemy, oh God!' At bar 37 the clamour subsides to allow the first soprano to be heard, conveying vital information for the audience in a recitative in strict time: 'from heathen Syria the proud Aminadab comes with floods of steel to Samaria, to overrun the kingdom.' As the strings re-enter and the music veers into B minor, the sopranos describe the panic, the slaughter, and the pessimism of the besieged citizens. The orchestra concludes with a D major ritornello, reworking the opening bars of the symphony. This accompanied duet is a striking and original way of setting the scene for the drama of human degradation that ensues. During the course of it Colonna explores a wide range of vocal forms, distributed among the soloists as shown in Table 8.1.

The orchestra participates in 18 of the 22 formal numbers. The 3 accompanied arias fall into the da capo mould: an orchestral introduction, section A in the tonic key, section B in a related key, and section

Ex. 8.1

Ex. 8.1 *(cont.)*

Ex. 8.1 *(cont.)*

per-bo   con tor-ren-te d'ac - cia - ro   vien di Sa-ma - ria

(from heathen Syria the proud Aminadab comes with floods of steel to Samaria)

TABLE 8.1. *Distribution of Numbers in* La profezia d'Eliseo

	W. 1 S.	W. 2 S.	Captain S.	Elisha A.	Joram B.
Continuo arias (2)	1	0	0	0	1
Continuo arias with coda (13)	3	3	3	4	0
Accompanied arias (3)	1	0	0	0	2
Duet with continuo (1)	1	1	0	0	0
Accompanied duet (1)	1	1	0	0	0
Duet arioso (1)	1	1	0	0	0
Accompanied chorus *a* 5 (1)	1	1	1	1	1

A in the tonic (an exact repeat). The first accompanied aria, King Joram's 'Ò corona, ò scettro, ò soglio' in Scene 2, is remarkable for its chromatic intensity, made more poignant by shifts in the texture from one instrumental group to the other (Ex. 8.2).

At the beginning of Part II, Woman 1, determined to protect her son's life, sings to him the aria 'Viscere care e belle'. As it is a continuo aria with orchestral coda, Colonna's favourite type of aria, it will serve to illustrate how he draws material from the aria (basically a two-part invention) to create an eight-part invention for the instruments. The sentiments expressed by the mother are affectionate and protective:

> Viscere care e belle
> Sì, sì, vivete, sì.
> Cadono in rie procelle,
> Sciolte dal ciel le stelle,
> Pria ch'io vi tolga il dì.
> Viscere care e belle
> Sì, sì, vivete, sì.

(Dear and lovely heart, yes, yes, you must live. May the stars released from the sky fall in evil storms, before I deprive you of life.)

Ex. 8.2

(O crown, o crown, o sceptre, o throne)

Colonna chooses the minor key to reflect her state of anxiety. His setting (Ex. 8.3) is simple and eloquent, with flowing coloratura on 'vivete' and 'tolga'. Though Neri's lyrics hint at a possible change of mood in the middle section, Colonna resists the idea and sets line 3 to a variant of line 1, thus maintaining a single mood. For the orchestral coda, he draws on motifs from both sections of the aria. The second orchestra has 'Viscere', 'Sì, sì', and a decorative figure from 'tolga'. The first orchestra has a variant of the 'Cadono' figure, but thereafter functions as a rhythmic counterbalance to its partner. Despite these

Ex. 8.3

Ex. 8.3 (*cont.*)

Ex. 8.3 (*cont.*)

(Dear and lovely heart yes, yes, you must live . . . before I deprive you of life . . .)

quotations of previously heard vocal motifs, the seven-bar ritornello is indeed a fresh surge of invention over a new basso continuo. Whilst the audience is savouring it, the demented Woman 2 makes her entry, eager to sink her teeth into the child.

The height of the women's quarrel over the boy (Ex. 8.4) presents quite a challenge to the composer: to create a sense of mounting hysteria in a recitative scene, 'Del famelico labbro'. He does it initially by overlapping the beginnings and ends of sentences over a slow-moving bass. The climax is reached when the continuo breaks into quaver movement, and then into a furious 3/4, in support of the arioso duet 'Estinto/Vivente'.

The finale of the oratorio features the Chorus singing 'Vittoria rimbombi, trionfi pietà'. All the assembled forces participate in a bright D major, the key with which the oratorio had begun. The

Ex. 8.4

Ex. 8.4 *(cont.)*

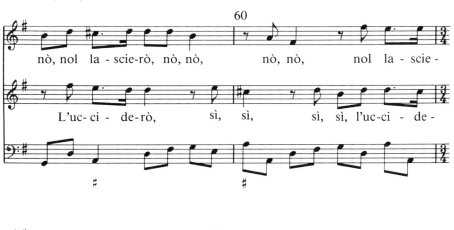

nò, nol la - scie-rò, nò, nò,   nò, nò,   nol la - scie-

L'uc- ci - de-rò,   sì, sì,   sì, sì, l'uc-ci - de-

- rò.   Vi - ven - te   vi -

- rò.   Es - tin - to in pas - to, per sem - pre in

- ven - te   per sem - pre l'ha - vrò.

pas - to, per sem - pre, per sem - pre l'ha - vrò.

(Fury. ‖ A beggarly thought. ‖ Wise advice. ‖ Give up your ill-starred son. ‖ Stop! ‖ Let him go, worthless woman. ‖ Pitiless woman, what are you trying to do? You are crazy! ‖ You'll see. Cut open that chest. ‖ Shut up! ‖ I will slaughter him, yes, yes! ‖ No, no, I'll not give him up. ‖ I'll kill him, yes, yes! Dead in a meal for ever I'll have him. ‖ Living for ever I'll have him.)

finale is a solemn processional piece in 3/2, in ternary form. In the brief middle section, Woman 1 and Elisha sing the praises of the God of Israel in duet. Woman 2 replaces Elisha for the last phrase *a 2*, thereby confirming her return to sanity. As the sound of rapturous voices dies away, Colonna's double orchestra adds a parting fanfare.

9

1686–1702
Antonio Gianettini

Born in 1648 in the parish of San Leonardo in Fano, a small town near Pesaro on the Adriatic coast, Antonio Gianettini[1] moved to Venice at the age of 24 to join the cappella of San Marco as a bass singer and organist. He remained there for fourteen years, working alongside Cavalli, Monferrato, Legrenzi, and Partenio in the cathedral, and composing six operas for Venetian theatres. Though he had to leave the Most Serene Republic to gain the promotion he coveted, he remained a lover of all things Venetian until his death in Munich in 1721.

Gianettini left Venice for Modena on 1 May 1686; the duke sent a barge to collect his belongings. He was awarded a large stipend of L396 per month, in addition to which an annual sum of 115 *scudi* (i.e. L575) enabled him to live in a fine house on the Corso Canal Grande. Such marks of favour show that the duke was pleased to acquire his services. In the first place, the new maestro was a native of Fano, the home town of the Martinozzi family and the birthplace of the duke's mother, Duchess Laura, so a touch of sentiment may have influenced his appointment. Secondly, Gianettini was a musician of wide experience, particularly in the field of vocal music, which, as events were to prove, the duke wanted to strengthen in Modena. Thirdly, he was a man of energy and good character who would work harmoniously with the Theatines in San Vincenzo; he chose to be buried at the Theatine church in Munich. The high value placed on Gianettini's services was justified; he proved a most industrious and efficient maestro di cappella. Not only did he strengthen Modena's existing cultural links with musicians in Bologna and Rome, but also he used his Venetian connections to good effect. Every year in January/ February, he attended the Venice Carnival to recruit musicians for the oratorio season in Modena, or for the summer season of chamber music and divertissements at Sassuolo. From 1686 to 1689 vocal music at court flourished as never before, with thirty-two oratorios and four

---

[1] An informative monograph on Gianettini is E. J. Luin, *Antonio Gianettini e la musica a Modena* (Modena, 1931).

operas performed under Gianettini's management. The cappella had never been so active, and reached a peak of twenty-nine members in the spring of 1689.[2]

A few weeks before Gianettini arrived in Modena, Giardini was promoted from his post in the chancellery to become Secretary of the Chamber, the duke's private secretary. For years his advice had been valued by Francesco II (e.g. over Ferrari's appointment as maestro), but the death of Secretary Nardi on 30 March opened the way for Giardini to become the duke's right-hand man, vetting and editing his postbag, arranging his diary, fixing audiences with the duke, authorizing payments from the privy purse, etc. All matters to do with the Cappella Ducale fell within Giardini's jurisdiction, a situation which Gianettini was quick to grasp. The new maestro appreciated that the quickest way to solve a problem, or get approval for a venture of his own, was to write to Giardini and ask him to bring it to the duke's attention. As both men were excellent administrators, sharing high artistic aims, their partnership was a great success.

In the autumn of 1686 the duke's fortunes took a turn for the better. His cousin Cesare Ignazio returned from exile and his uncle Rinaldo was finally raised to the cardinalate by Pope Innocent XI on 2 September. In November Francesco left his duchy, accompanied by a large entourage of courtiers and officials (including Giardini), to visit his mother in Rome and sample at first hand the attractions of the city's religious and cultural life. While he was there, Rome celebrated the coronation of his sister and brother-in-law in a series of splendid concerts organized by Queen Christina's circle. Pasquini contributed a cantata for the occasion at the Palazzo Riario in February 1687 and, at the same venue, Corelli directed a huge orchestra of 150 players to mark the arrival of the new British ambassador, Lord Castlemaine. One slight disappointment marred the duke's visit: his mother's refusal to give up her self-imposed exile. After spending a few days in Naples, Francesco returned home on 25 February, in time to discover the capabilities of Gianettini as director of the new season of oratorios in San Carlo rotondo.

The 1687 season had a strong Roman flavour: six of the nine oratorios presented in Modena were composed by Roman musicians. Three were the work of Pasquini. In *I fatti di Mosè nel deserto* [38], based on Exodus 18, Moses (S.) and Joshua (A.), helped by a Narrator (T.), give an account to Jethro (B.) of their trials and triumphs in the wilderness. This setting of the fifth episode of *La vita di Mosè* by

---

[2] Added to the orchestra in 1689 were N. M. Ferri, G. B. Baldrini, Domenico Gabrielli, Marco Martini, and G. Massoni.

Pasquini was heard alongside revivals of two of his Roman oratorios, *S. Alessio* [36] and *Il martirio dei santi* [37]. The former contains a delightful sinfonia *a 10* in the Roman concerto style to depict the festivities at the saint's impending nuptials, the latter restates, through the triple martyrdom of Vito, Modesto, and Crescenzia, the belief that Christian martyrdom is a blessed fate.

The other Roman oratorios of the 1687 season were Melani's *Il sacrificio d'Abelle* [35], Lanciani's *S. Dimna* [34],[3] and Agostini's *Primo e secondo Miracoli di S. Antonio* [30]. The oratorios of Melani and Lanciani are stories of martyrdom within the family circle: Abel slain by his jealous brother, and the Irish princess Dimna stabbed by her pagan father for refusing to share his bed. As *S. Dimna* was dedicated to Duchess Laura, it seems likely that it was performed in Rome during the duke's stay (no Roman libretto survives), prior to its transfer to Modena. Agostini's setting of the two miracles of St Antony is on a grand scale, befitting the supernatural subject-matter. The first miracle ends with an elaborate madrigal chorus *a 5*, accompanied by a five-part string ensemble (the counterpoint runs to nine real parts). The second contains a stirring sinfonia *a 12*, scored for four groups of contrasting timbre: two violins and bassoon, two cornetts and trombone, two violas and violone, and two trumpets and timpani. The libretto of the miracles has a third episode which, apparently, Agostini never set to music. Under the title *Amore alla catene* [32], Gianettini made a splendid setting of the third miracle—it ends with a concerted madrigal chorus in eight real parts matching Agostini's conclusion to the first miracle—so that the whole cycle could be performed for the Congregazione di San Carlo in 1687.

Gianettini's skill as a contrapuntist was also in evidence in *L'Huomo in bivio* [33]. It is a morality play of the Venetian sort, in which Man (A.) is tempted by a Demon (B.) and a varied assortment of choral groups: Passions (S. T. B.), Vices (S. A. T.) and Infernal Spirits (S. A. T. B.), all urging him to take the road with them. Persuading him to go the other way are an Angel (S.), a Chorus of Spheres (S. A. T. B.) and a Chorus of Angels (S. A. T. B.). Battle royal is joined at the crossroads (*bivio*). It ends with a duet when the triumphant Angel brings the Demon to his knees in remorse, and the Chorus of Angels comments: 'Blessed is the man who follows heavenly teaching, for it puts him on the right road.'

If some of the Modenese felt a little intimidated by the number of foreign works heard in their oratory in 1687, they had the consolation

---

[3] For a modern facs. edn. of *S. Dimna*, see J. L. Johnson and H. E. Smither (edn.), *The Italian Oratorio, 1650–1800* (Garland Press facs. repr., New York, 1986), vi.

of knowing that one at least was a product of local genius, *La vittoria di Davide contra Golia* [31]. The composer was Giovanni Bononcini, eldest son of Giovanni Maria Bononcini, born in Modena in 1670.

Giovanni was only 7 years old when his father died. He had a remarkable aptitude for music and his stepmother, in accordance with her husband's dying wish, sent the lad to Bologna to study under Colonna at San Petronio. At 15 he published his first set of trio sonatas in Bologna and dedicated them to Duke Francesco. Five more instrumental sets followed in rapid succession. At 17 he was admitted to the Accademia Filarmonica, joined the regular cappella of San Petronio, and at the same time directed the cappella at the ancient church of San Giovanni in Monte in Bologna. The indefatigable youth turned to vocal composition in 1687 with his first oratorio, *La vittoria di Davide*, dedicating this also to the duke as proof that he was making good progress in his chosen profession. That Francesco was interested in his progress—he could well have paid for his tuition—is clear from the dedicatory address of Bononcini's second oratorio, *Giosue* (1688). He says that his unceasing labours under the skilled teaching of Colonna have given him the strength of an Achilles. Brimming with confidence, Bononcini promoted his first two oratorios at Bologna and Modena as soon as they were finished.

Unfortunately the score of *Davide* is lost. From the libretto we can observe that the oratorio follows the Bible story with little embellishment. Part I begins with Goliath calling the Philistines to arms and ends with a fiery duet in which Saul and an Israelite Captain urge David to give the heathen his deserts. The confrontation between David and Goliath is cast in an agitated recitative scene, the giant expiring with the cry of 'Traditor!' on his lips. As Bononcini had already published a set of trumpet sonatas (Opus 3), one can guess that the battle scenes in *Davide* were rendered in typical Bolognese fashion. The texts of the formal musical numbers, however, accord with the latest poetic vogue: twelve arias and a duet are in da capo form.

In mid-July 1687 Duchess Laura died in Rome. The tone of her last letter to her son, sent from Frascati on 4 June, shows that his visit had brought a measure of reconciliation, though not enough to tempt her back to Modena:

I see in your letter, full of kindness towards me, the care you have for my health, for which I am much obliged. But I assure you I am presently in no fit state either to make the journey to England, or even as far as Modena. Even a small journey does me harm, and also, it would be pointless for me to travel with the summer still advancing, the heat and the dust which one cannot avoid being very bad for me. Yet I cannot end without thanking you [for the invitation] and begging you to be persuaded of the tender and cordial affection I have for you. Meanwhile I remain ever your most affectionate mother.

LAURA, Duchess of Modena[4]

The pope sent his condolences to the queen of England on 19 July, the court at Modena went into mourning, and by the end of the month de Grandis was writing to Giardini from the Santa Casa at Loreto, 'For the death of the Most Serene Mother of His Serene Highness I offered Masses in the Sanctuary for her soul. God has her in Glory.'[5] Laura's state funeral in Modena occurred a year later on 3 August 1688. It was conducted among the Este monuments she had erected in Sant'Agostino. Her old confessor Padre Garimberti gave the oration.

The duke was very distressed by his mother's death. To add to his troubles, he fell seriously ill in the spring of 1688. On 24 April de Grandis reported that all Loreto was talking about the duke's infirmity and praying for his recovery. He also added a marginal note: 'I understand that some superb oratorios have been performed this Lent, but it is no miracle to me where there are so many virtuosi'. De Grandis's information was absolutely correct; the oratorio season of 1688 in Modena was of the highest quality. No other venue in Italy could boast a better season than this until the *filippini* in Florence contracted oratorio fever in 1693.[6]

Gianettini's own contribution was a setting of *La vita di Mosè, 6: La creatione de' magistrati* [41].[7] It was a much more sophisticated piece than the two oratorios he had offered in his first season as maestro, and the only one in 1688 originating in Modena. The other nine came from Rome, Bologna, Ferrara, and Venice.

From the duke's collection of Stradella manuscripts Gianettini extracted two of his Roman oratorios, *S. Giovanni Battista* [48] of 1675 and *S. Pelagia* [47] of *c*.1673. Both oratorios have attracted a good deal of critical attention over the years.[8] In directing performances of them in Modena, Gianettini had the advantage of being able to consult Siface about the complexities of *S. Giovanni Battista*. The

---

[4] The Italian text of the letter is in C. de Cavelli, *Les Derniers Stuarts à Saint-Germain-er-Laye* (Paris, 1871), ii, doc. 194.

[5] The letter is in *I-MOs* MM, Busta 1A.

[6] For details of the oratorio repertory in Florence, see R. Lustig, 'Saggio bibliografico degli oratorii stampati a Firenze dal 1690 al 1725', *Note d'archivio per la storia musicale*, 14 (1937); J. W. Hill, 'Oratory Music in Florence, II', *Acta Musicologica*, 51 (1979), 246–67; and id., 'Oratory Music in Florence, III', *Acta Musicologica*, 58 (1986), 129–79.

[7] Details of the oratorio are in Ch. 10.

[8] For a recent appraisal of *S. Giovanni Battista* see H. E. Smither, *A History of the Oratorio*, i (Chapel Hill, NC, 1977), 316–27. *Santa Pelagia* is analysed in Niall O'Loughlin, 'Stradella's Santa Pelagia', *Musical Times* (May 1981), 297–300.

duke's castrato had sung the role of Herodiade for the composer some thirteen years earlier. Of more recent date was Lulier's *S. Maria Maddalena dei pazzi* [42]. First performed at the Palazzo Panfilio on the Corso in Rome in October 1687, the oratorio was then sent to Modena for production in 1688. Likewise, in 1689 Lulier's *S. Beatrice d'Este* [58] was produced, first in Rome, shortly afterwards in Modena, reinforcing the cultural links between the two princely households forged by Pasquini's *S. Agnese* back in 1685. Lulier's setting of *S. Maria Maddalena dei pazzi* contains some astonishing bravura passages for the heroine (a castrato role) and for the solo violinist who accompanies two of the saint's arias.[9]

The contribution from Bologna in 1688 consisted of three oratorios on Old Testament subject-matter. The story of the escape of Lot and his daughters from Sodom is recounted in Millanta's *Loth* [43]. In Colonna's *La caduta di Gierusalemme* [40],[10] Jerusalem is destroyed by Nebuchadnezzar, king of Babylon. The composer, though constrained by his text (Bergamori again) to provide pages of serviceable recitative for long speeches, springs into lively action whenever an aria or ensemble is required. The battle for Jerusalem reaches its peak with a brilliant concerted quintet, 'Grandinate le ferite', a technical *tour de force* by the maestro of San Petronio. There was also plenty of technical bravado in Bononcini's *Giosue* [39], depicting the fall of Jericho. Colonna's disciple had evidently learned much from his master about the handling of instrumental resources, about coloratura and about chromatic harmony. But already in his second oratorio Bononcini is developing a personal style more akin to that of Perti and the Romans than to Colonna's. Richer harmonies, simpler and more affecting melodies, and delicate instrumental coloration, particularly in arias with solo obbligatos, are the modern features of his style which were to pave the way to international acclaim. Trumpets are vital to the story of Joshua's triumph at Jericho. When they are mentioned in the text, Bononcini reverts to the natural trumpet key of D major, so that they might be employed, but makes no specific demand for them in his score. In this, he is following Colonna's practice rather than Perti's.

The most unusual oratorio produced in Modena in 1688 was Pallavicino's *Il trionfo della castità* [46]. It was scored for five sopranos, two altos, and a 'musico' (castrato) who sang only the final aria 'con tromba'. This strange assortment of high voices, matched by an equally strange list of characters (female saints, Rhenish princes,

---

[9] A facs. ed. by Johnson and Smither is in *Italian Oratorio*, vi.
[10] Ibid. v.

allegorical figures—one disguised as a nurse) points to the strong probability that the oratorio was composed as a showcase for the talents of Pallavicino's singing students at the Conservatorio degl'Incurabili in Venice, where he served as maestro di cappella from 1674 to 1685. With the exception of the final aria, the technical demands on the singers are modest, as are the demands on the string ensemble *a 4* which provides the accompaniment.[11] For the girls of the Incurabili, *Il trionfo della castità* must have been a welcome diversion from the rigours of the schoolroom, and its performance an opportunity for the citizens of Venice to sample the talents of potential prima donnas. As Pallavicino died in Dresden in January 1688, the instigator of the Modenese performance was probably the librettist, Giannini. For many years Giannini had kept the Modenese court informed about Venetian cultural affairs; he had helped to recruit singers for *Il principe corsaro*, the first opera of Francesco II's reign in 1674.[12] Whether or not Venetian girls sang the oratorio in Modena is an open question. They might have done so, for the libretto printed by Soliani is not included in the firm's collection of oratorios[13] performed in San Carlo rotondo (a male preserve). Female performers would account for the exclusion. We know that the performance was sponsored by the duke because the State Archive has a bill from Giuseppe Colombi, showing that he was paid L40 for copying the score and parts on 9 June 1688.[14] The court connection is indisputable; the venue and performers of *Il trionfo della castità* remain an intriguing but unsolved mystery.

Two oratorios of Padre Palermino of Ferrara were produced at San Carlo rotondo in 1688: *Sansone* [45][15] and *S. Rosalia* [44]. In the latter, the saint (S.) is accosted by three allegorical figures: Penitence (S.), Ambition (A.) and Passion (T.). As she opts to take the hard road of penitence she is attacked by Lucifer (B.). Only by invoking the help of the Virgin Mary (S.) does she fend off the attack and achieve contentment. This sequence of events occurs in Part II. Part I is a prologue (in all but name) for the three allegorical figures, who dispute each other's claims to primacy in a series of brilliant arias, duets,

---

[11] Both Arnold Schering in *Geschichte des Oratoriums* (Leipzig, 1911) and D. and E. Arnold in *The Oratorio in Venice* (RMA Monographs, 2; London, 1986) have commented on the moderate technical demands made by Pallavicino. There is no Venetian libretto extant, but the internal evidence strongly suggests a première at a Venetian conservatory.

[12] A long letter from Giannini to Tagliavini, dated 17 Feb. 1674, is extant in *I-MOs* MM, Busta 1B.

[13] Eight oratorios performed in San Carlo rotondo in 1688 appeared in Soliani's *Raccolta d'Oratorii 3*. The fact that Pallavicino's oratorio was not included in the collection, though published by the same house in 1688, is intriguing. Performance at a different venue in Modena would be the simplest explanation of the exclusion.

[14] Copyists' bills are in *I-MOs* MM, Busta 3.

[15] For details of the oratorio see Ch. 11.

and trios. Such scenes were stock-in-trade for Venetian opera composers, though few carried them off with such panache as Palermino
demonstrates in *S. Rosalia*.

On 10 June 1688 the duke's sister gave birth to a son, a Catholic
heir to the throne of England. Any rejoicing in this event was purely
for the family and well-wishers; in political terms it spelled disaster
for the Stuarts. The Protestant Parliament in London, irked by James
II's high-handedness in offering preferment to Catholics in contraven-
tion of English law, and dismayed at the prospect of an heir continuing to do likewise for the foreseeable future, invited William of
Orange and his wife Mary (James's eldest daughter) to re-establish a
Protestant monarchy in England. A bloodless revolution ensued
which forced Maria Beatrice and her infant son to fly to the protective
custody of Louis XIV in December. It was a rude awakening to
political reality for the Estensi, though for years after the expulsion
they nurtured hopes of a restoration. For loyal supporters of the
dynasty like de Grandis, hope centred upon an act of divine intervention. In a letter to Giardini from Loreto dated 5 February 1689, he
wrote:

The Canons here are asking if they might have from Modena a full account
of the great disgrace that has befallen their Majesties of England. They hear
at present that, with the most miraculous Divine Help, they are safe, and
being magnificently treated by the Most Christian King [Louis XIV]. But
because some say one thing, some another, would you do me the honour of
granting their request

He ends with the pledge that the Santa Casa will use every means
(masses, prayers, etc.) to re-establish the great common cause. To the
confusion of all heretics, he hopes to see shortly a heavenly miracle.[16]

One can well imagine that the feelings and hopes expressed by de
Grandis in this letter were shared by the members of the Congregazione
di San Carlo in Modena. Their response to the dynastic misfortune
was to rally to the common cause and produce thirteen oratorios in
the spring of 1689. As the prayers of the congregation centred upon
the plight of the queen of England, so the subject-matter of many
of their Lenten oratorios dealt with similar predicaments found in
Scripture or the hagiographies. Such stories gave ample scope to the
preachers who delivered sermons in the oratory, to draw out the
parallels and offer moral support to the dynasty.

The cult of the Virgin Mary, who suffered grievously for and with
her son, was an obvious source of spiritual consolation for a society
bearing her name. Ferri's *La Vergine annonciata* [51], Marco Martini's

<hr>

[16] The letter is in *I-MOs* MM, Busta 1A.

La verginità trionfante [59], and Pasquini's *La sete di Cristo* [60] present vivid images of Mary's innocence, faithfulness, and courage under stress, for pious contemplation. In addition, Gianotti's *Il constituto di Cristo* [57], depicting Christ's trial before Pilate and St Peter's betrayal, points the moral that to be faithful to the truth of God entails suffering and, *in extremis*, martyrdom. If we turn to the group of oratorios based on the lives of the saints (excluding Gabrielli's *S. Sigismondo* [53], composed for the Oratorians in Bologna two years earlier, and Lulier's *S. Beatrice d'Este* [58], composed in Rome[17]) we discover two stories featuring mothers and sons: Gabrielli's *S. Felicità* [52] and Gianettini's *La conversione della Beata Margherita* [55]; and another tale of an innocent virgin meeting an untimely death with composure, Gianettini's *Il martirio di S. Giustina* [56]. The case for thinking that these seven oratorios constitute a reaction to the plight of the Stuarts is strengthened by the fact that the poets who wrote six of them (Pasquini's *La sete di Cristo* being the exception) were literati of the Modenese court circle.

Completing the list of productions in 1689 were four oratorios based on Old Testament stories: Vitali's *Giona* [61],[18] Bassani's *Giona* [50], Gaffi's *Abigaille* [54], and Alessandri's *Bersabea* [49]. The last two were oratorios about the wives of King David, a sequence rounded off in the following season when Gaffi, maestro to Cardinal Rinaldo d'Este in Rome, sent a setting of *Micole* [64] to Modena. As the duke of Modena was 30, and still unmarried, tales of David's connubial bliss were perhaps timely reminders that a monarch owed it to his subjects to take a wife and beget an heir.[19]

To produce thirteen oratorios in one season put considerable pressure on the Cappella Ducale, particularly upon Frignani's department of copyists. To cope with the load, Frignani had to call upon five assistants (Giovanni Braida, Andrea Sarti, Domenico Giannini, Giuseppe Colombi, and Domenico Leporatti) to keep on schedule.[20] But feverish activity does not guarantee high quality. Of the scores surviving from 1689, only those by Lulier, Pasquini, and Vitali rise above a workaday level.

Since the appointments in 1686 of Giardini as paymaster of the cappella, and Gianettini as its maestro, the cost of music productions (oratorios and operas) at court had soared. In the summer of 1689 at Sassuolo, the duke decided to reform his cappella, and began the

---

[17] An eye-witness account of the première is printed in 'Studi Corelliani', *Rivista italiana di musiclogia*, 3 (1972), 116.
[18] For details of the oratorio see Ch. 12.
[19] Francesco's sister had been urging him to get married since 1681.
[20] Copyists' bills are in *I-MOs* MM, Busta 3.

process by dismissing eight instrumentalists. There are several documents in the State Archive relating to the reform,[21] revealing that it was not undertaken without careful thought being given to the balance of the cappella and the competence of its members, but two documents are of special importance: an official minute to Gianettini dated 5 July, drafted by Giardini, and the maestro's letter of response sent from Modena on 7 July. The minute is entitled, 'Instrumentalists who were salaried by Duke Francesco II'. It reads:

His Most Serene Prince the Duke, having resolved to reform the Body of his instruments and remove from the payment register the below-mentioned instrumentalists who, to the prudence of His Serene Highness seem superfluous and unnecessary for the service of his Cappella, commands me to write to you to ask you to make known to each of them the resolution taken by His Serene Highness. They are: Casanova, Capiluppi, Barbieri, Ciocchi, P. Ascanio, P. Severi, Ipp[olito] Bellini, D. G. Baldrini. His Serene Highness has given orders in the Chamber to remove the same from the salary register.

The minute offers Gianettini better news with the announcement that the duke has taken into service two other instrumentalists, Pincelli and Frangiolli, and is considering appointing a violettist from Venice if he is free to come to Modena. Gianettini's reply, posted two days later, states that he has done his duty and dismissed the eight instrumentalists. After discussing the new appointees, he expresses his own satisfaction with the reforms.

Now we are in a position to make a good repair of the Cappella, to perfect it where there were shortcomings . . . We must have the diligence to choose active men of proficiency, because the service of this Cappella needs men who know their trade and practice it well, for we have a Prince of excellent taste who needs the best men to serve him.

What was happening behind the scenes, concealed by the maestro's optimistic tone, was the dismantling of the court orchestra. In August 1693 three more instrumentalists were dismissed. As the elder Vitali had died in post in November 1692, the size of the salaried orchestra had been reduced from twenty in the spring of 1689 to four in the autumn of 1694. Only three violinists, Colombi, Tomaso Vitali, and Allemani, and the organist Bratti, remained on the payroll. In the last four years of Francesco's reign the administration of the court orchestra returned to the system of occasional hire that had pertained in the frugal years of the regency period. Meanwhile, during the same period, the number of singers remained stable at eight.

Though the number of oratorio productions fell to an average of

---

[21] In *I-MOs* MM, Busta 1B and Busta 2.

five per season in the period 1690–4, the repertory contained many new delights. Bononcini's third oratorio, *La Maddalena à piedi di Cristo* [62] was performed in 1690. In the following season, *La vita di Mosè* was completed with Gianettini's *Dio sul Sinai* [68] and Alessandro Melani's *Lo scisma del sacerdozio* [69]. The duke's marriage to his cousin Margherita Farnese of Parma in 1692 was an occasion graced by Monari's festive oratorio, *Il fasto depresso* [72] and Pistocchi's tuneful *Il martirio di S. Adriano* [74].[22] One of the last oratorios heard by Francesco was Vinacesi's sparkling setting of *Susanna* [87] in 1694.[23] Most of the oratorios of this period were progressive in form and style. Bononcini and Pistocchi, for example, channelled their energy into the creation of long da capo structures for arias, ensembles, and sinfonias, and also into the invention of expressive obbligato accompaniments for solo instruments. The elaboration of arias, in fact, considerably lengthened the time taken to perform oratorios; Bononcini's *Maddalena*, with its forty-three arias, runs for over two hours.

Ill-health and anxiety about affairs of State clouded the last years of Francesco's reign. The marriage, which should have rekindled hope, proved childless. The expense of it depleted the state coffers to such an extent that musicians in the cappella were unpaid between November 1692 and February 1694. In organizing seasons of oratorio, Gianettini was obliged, for the first time, to trim expenditure by recycling works from previous seasons, thus saving on rehearsal time and copyist's bills.

The vitality of the oratorio tradition in Modena continued to decline in the late 1690s. Many of the musicians whose talents had sustained it in the past died in the 1690s: Vitali in 1692, Colombi in 1694, Colonna in 1695, and Siface (assassinated) in 1697. Young composers like Bononcini and Pistocchi, who were capable of bringing new life to the genre, left their native soil to seek fame and fortune elsewhere. The greatest blow, however, came in September 1694 when Duke Francesco II died from an acute attack of gout. His successor Rinaldo, though a prince of the Church, had neither the religious nor the musical sensibility of his nephew. He regarded the oratory as a place to be frequented only on State occasions. The visit of Cardinal Grimani in 1698 was one such occasion; Gianettini's *S. Agostino* [98] was performed for the honoured guest. In 1699, an imperial wedding, the baptism of the crown prince of Modena, and the death of Rinaldo's mother Duchess Lucrezia Barberini d'Este, were each marked by performances of oratorios. State occasions in the

[22] For details of the oratorio see Ch. 13.
[23] For details of the oratorio see Ch. 14.

past had brought forth new works, specially commissioned, but in Rinaldo's reign almost all the oratorios were old ones, dusted off by Gianettini at the duke's bidding. Only two new oratorios were produced in the first seven years of his reign: Bassani's *La morte delusa* [91] and Gianettini's *S. Agostino*.

To give credit where it is due, Rinaldo did strengthen the Cappella Ducale. It reached a complement of two maestri, ten singers (including a female soprano, Margarita Salicoli), ten instrumentalists, and five buglers by the end of the century. But the duke's lack of interest in religious music knocked the heart out of Modena's oratorio tradition. It was moribund long before the State itself collapsed with the French invasion of August 1702.

Gianettini's
La creatione de' magistrati (1688)

La creatione de' magistrati, an oratorio in two parts (*Vita di Mosè*, 6)
Music by Antonio Gianettini
Libretto by Giovanni Battista Giardini
Characters: Sefora (Zipporah, wife of Moses) S.
               Moisè (Moses)   A.
               Getro (Jethro, Zipporah's father) B.
Instrumental accompaniment *a 4*
First performed in the oratory of San Carlo rotondo, Modena, on 4 April 1688.

Antonio Gianettini (b. Fano, 1648; d. Munich, 1721) came to Modena as court maestro in May 1686, and immediately struck up a close friendship with Giardini, the duke's private secretary. It was a natural development, therefore, that Giardini should invite the new maestro to set to music episodes 6 and 7 of *La vita di Mosè*; the present oratorio and *Dio sul Sinai* (1691).

The Libretto

By the sixth episode of *La vita di Mosè*, Giardini had reached Exodus 18, and a clear pattern had evolved in his handling of the epic. Each episode explored a particular facet of Moses' character. In the first two, set to music by de Grandis, we had seen the baby Moses, an innocent hostage to fortune but blessed by God, and the bridegroom Moses, the gallant protector and consort of Zipporah in Midian. In episodes 3 and 4 Moses as God's spokesman in Egypt and Israel's saviour at the Red Sea had been revealed. Moses' leadership of Israel in the wilderness had been the subject of episode 5. Now, in episode 6, Moses' role as a judge of Israel is under scrutiny.

The reunion of Moses with his wife and father-in-law, described in Exodus 18: 6–27, forms the basis of Giardini's oratorio. As the encounter is a personal one, and there is no action other than Jethro giving Moses sound advice, Giardini dispenses with a narrator and restricts the characters to three, Moses' family circle. It is worth noting, in passing, that Jethro, the heathen priest of Midian, had already appeared in episodes 2 and 5 of the cycle, and his daughter in episode 2.

On pages 7 and 8 of the libretto, Giardini dedicates his oratorio to Duke Francesco II. He begins:

In the first ages of the world, when human politics was new, and stuttering, in need of tuition, the Holy Spirit instructed Moses through the mouth of Jethro. At the time when he was discussing the creation of magistrates, he warned him that he should elect subjects of authority, fearful of God, truthful and not covetous. 'Elige', he said, 'viros potentes de Populo, timentes Deum, in quibus sit veritas, et qui oderint avaritiam' [Exodus 18: 21]; because, in fact, these are the four wheels which keep steady the Triumphal Coach of Justice, and the four fundamental pillars upon which rests the machine of government

It was a shrewd move on the author's part to make an apology for Jethro's heathen origins by claiming that these were primitive times, and to buttress his case by an apt quotation from scripture. Giardini, as we have seen before in his handling of *Susanna* (Chapter 4), knew that the best way to secure freedom of artistic expression for himself was to assure the ecclesiastical censors of his orthodoxy.

The dramatic structure of the oratorio is as follows:

Sinfonia
Part I:
Scene 1   Moses, Zipporah: Early in the morning, Moses takes leave of his wife to attend to his onerous duties as the sole judge of the Israelites.
Scene 2   Jethro, Moses, Zipporah: Jethro asks where he is going at such an early hour. Moses confesses that his duties occupy him from dawn to dusk. Jethro and Zipporah urge him not to overtax himself. Jethro advises Moses to appoint magistrates to deal with lesser affairs of state. He tells him to be on the look-out for lying counsellors, greedy ministers, and members of tribunals who take bribes.
Part II:
Scene 3   Jethro, Zipporah: Jethro announces his departure, to the great distress of Zipporah. He counsels his daughter to comfort her husband, dress with modesty, and bring up her sons to be virtuous.
Scene 4   Moses, Jethro: Moses enters, a happy man. He thanks Jethro for his sound advice and explains the newly estab-

lished system of judicial delegation. Jethro is pleased, but warns Moses to be watchful.

Scene 5   Jethro, Moses, Zipporah: Jethro prepares to depart. A mortified Zipporah is comforted by Moses. Jethro gives them his blessing. In a trio they sing their farewells.

The domestic scenes (1, 3, and 5) are products of Giardini's imagination, providing a pleasant, homely framework for the legal discussions of Scenes 2 and 4, and, at the same time, fleshing out the character of Zipporah.

Giardini is scathing about corruption in high places. The tone of Jethro's attack on greedy ministers at the end of Scene 2 is as bitter as the Narrator's attack on greedy lawyers in *Susanna* (Chapter 4).

JETHRO:	Mà guai, si instinto avaro
	Nel ministro prevaglia. Astrea depresse
	Da Tiranno interesse
	Piangerà le sue leggi.
	Vedrà le sue bilancie
	Destinate à vil uso,
	Pesar l'oro e non l'opre, e del suo brando
	Vedrà guaste le tempre el'filo ottuso.

(But woe betide, if a greedy instinct prevails in a minister. Justice, depressed by tyrannical interests, will weep for her laws. She will see her scales put to base use, weighing gold rather than deeds, and she will see the blade of her sword spoilt, its edge blunt.)

As a long-serving clerk in the chancellery, Giardini had considerable experience of court administration on which to draw when overtaken by a fit of righteous indignation.

The most important parts of the dramatization occur in passages of recitative. The manner of speech is highly rhetorical as Jethro, in Scene 2, gives Moses his spiritual instructions (cf. the Dedication).

JETHRO:	Odi, odi, ò figlio.
	Se m'ami, odi. Odi i miei detti.
	Frà le Turbe seguaci
	Eleggi tu soggetti
	E provetti e capaci
	A reger teco il peso degl'affari.
	E sia lor cura
	Le materie più grave
	Riferir al tuo soglio.
	E spedir soli quelle meno importanti
	De la vita del Regno,
	De la mente del Rè, braccia e sostegno.

(Listen, listen, O son. If you love me, listen. Hear my advice. From among your host of followers appoint experienced and capable subjects to judge with you the weight of cases. And be it their responsibility to refer to your throne the more serious cases and to deal only with those of lesser importance to the life of the kingdom; a brace and support to the king's intelligence.)

The subject-matter of the arias and ensembles in the oratorio ranges from moral philosophy and politics to direct expressions of family affection. In Scene 2, Moses' aria 'Vera imago de Regnanti' compares the onerous duties of kings to the never-ending journeyings of the sun, while Jethro's 'Peste rea del Consiglio' pours scorn upon the deceitfulness of counsellors. Personal affections dominate the final scene. Zipporah, crestfallen at her father's departure, sings an aria *affettuosa*:

> Se à figlia che prega
> Dà un padre si niega
> Si poca mercè,
> O manca l'Amor
> O tanto non merta
> L'afflito mio cor.

(If a father responds so negatively with such little pity to a daughter who prays, either he lacks love or my heart does not merit such torment.)

Moses responds to her anguish by urging her to stay for his sake:

> Qui t'arresta, o caro ben.
> Che se tratti di partir
> Il martir
> Fà ritorno in questo sen.

(Stay here, my dear love. If you decide to leave, the torment will return to my heart.)

Jethro's blessing (in recitative) ushers in a tearful leave-taking:

JETHRO:	Addio, figli.
ZIPPORAH/MOSES:	Padre, addio.
JETHRO:	Io parto.
ZIPPORAH/MOSES:	Tu parti.
JETHRO:	Qui resta l'mio core.
ZIPPORAH/MOSES:	Vien teco l'mio core

(Farewell, my children. ‖ Father, farewell. ‖ I am departing. ‖ You are departing. ‖ My heart stays here. ‖ My heart goes with you.)

Such sentiments were the stock-in-trade of Venetian opera librettists, and thus familiar ground for a composer like Gianettini, with fourteen years of service in Venice behind him.

The Score

The score of *La creatione de' magistrati* (*I-MOe* Mus. F. 501) is a beautiful presentation copy, with a decorated title-page and ornamental capitals. It was produced by Antonio Frignani, copyist in the cappella since 1677. The scores of Ferrari's *Sansone* (1680), Colonna's *Il transito di San Gioseppe* (1681), de Grandis's *Il nascimento di Mosè* (1682), and Vitali's *Giona* (1689) were also his handiwork. The oratorio is scored for three solo voices, accompanied by strings *a 4* throughout. The basso continuo is amply figured in discordant passages.

Gianettini was evidently eager to impress. In the first twelve pages of the score he presents a noble sinfonia in C minor containing a four-part fugue, a ground-bass aria for Moses lasting 112 bars, an aria di bravura for Zipporah, and a graceful love-duet in the Venetian manner. The music is tuneful, sure-footed, and well proportioned, exactly what one would expect from the court's efficient new maestro.

The opening sinfonia is in two sections: 'grave' and 'allegro' (Ex. 10.1). After playing it, the orchestra is required to participate in all arias and ensembles except the final trio, where its abstention is dramatically effective. Its chief function is to round off the singing with a substantial ritornello; long orchestral codas are a hallmark of all Gianettini's dramatic compositions.

Ex. 10.1

Ex. 10.1 *(cont.)*

Of the twenty-one formal numbers in the oratorio, sixteen are da capo structures, the rest are through-composed. Two arias allotted to Moses are constructed over a modulating ostinato. The range of settings, and their distribution between the singers is as shown in Table 10.1.

TABLE 10.1. Distribution of Numbers in *La creatione de' Magistrati*

	Zipporah S.	Moses A.	Jethro B.
Continuo arias with coda (12)	4	4	4
Continuo arias with introd. and coda (3)	1	0	2
Accompanied arias (3)	1	2	0
Duet with coda (1)	1	1	0
Trio with introduction (1)	1	1	1
Continuo trio (1)	1	1	1

The ground bass arias 'Sallo il Ciel' in Scene 1 and 'Vera imago de Regnanti' in Scene 2 are both expressive of weariness. In the first, Moses is rising at dawn to take up his onerous duties, and in the second, he is comparing his task to the endless journey of the sun. In both cases, therefore, the mechanical repetitions of ostinatos are appropriate symbolic devices. 'Sallo il Ciel' (Ex. 10.2) has an eight-bar ground, played fourteen times: five in C minor, three in G minor, and six in C minor. This tonal format accommodates an expressive da capo aria with orchestral coda. With great expertise, Gianettini integrates the voice and continuo by using an imitative head-motif, and takes care to bring his vocal phrases to an end well clear of the eight-bar junctions in the ostinato.

Ex. 10.2

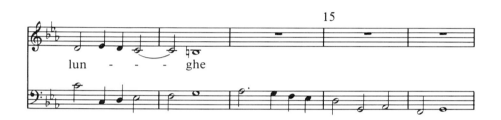

(The day is dawning; how much longer will it delay?)

By the time we reach 'Vera imago' (Ex. 10.3), Moses is so weary
that Gianettini deliberately lets the progress of his halting, nine-bar
ostinato govern everything: vocal phrasing and orchestral ritornello. It
makes brief excursions from the tonic key, but the sectional mode of
construction, relieved only by a couple of lethargic vocal phrases
(unaccompanied), makes a tedious impression, especially when Moses
embarks on a second stanza. Zipporah's comment (in recitative) at the
end of the aria must have brought a knowing smile to many faces in
the audience: 'After long hours of labour, Heaven concedes a decent
rest'! If Moses fell asleep at this point in the drama, then the extraordinary opening of Jethro's next speech, his repeated cries of 'Listen,
listen, O son', would fall into place as a bold attempt to wake him up

Ex. 10.3

(A true image of rulers, the sun revolves through the winds and spheres)

to hear the instructions of the Holy Spirit. To suggest that Gianettini and Giardini were playing upon the tolerance of their audience may seem far-fetched, but the concept of a 'deliberately boring' aria on a ground bass, devized for purposes of dramatic characterization, had been explored by Stradella in setting the Counsellor's aria 'Anco in Cielo', in *S. Giovanni Battista* (Rome, 1675). Furthermore, it should be noted that Stradella's masterpiece was revived by Gianettini in Modena in 1688, the very same season that *La creatione de' magistrati* came to light. A prototype for 'Vera imago' was to hand.

Zipporah's distress at the impending departure of her father in Scene 3 gives Gianettini an opportunity to compose some highly charged music. In the recitative 'affettuoso' (Ex. 10.4), chromatic

Ex. 10.4

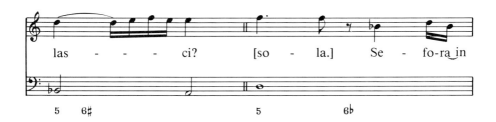

Ex. 10.4 (*cont.*)

ta al duo - lo.

(And will it be true, oh God, that you would leave me? . . . [alone.] Sefora in tears, abandoned to grief.)

harmonies, sighs, and realistic sobs on 'abandonata' lead to a Phrygian cadence on A. The pathetic aria that ensues (Ex. 10.5) startles the ear by commencing on a 6–3 chord of E flat major, at a tritone's remove from the end of the recitative. It eventually settles in D minor and, in the tenth bar, the strings begin to add their doleful harmonies to the heroine's agonized complaint.

The ensembles in the oratorio are strategically placed: a love-duet for Moses and Zipporah at the end of Scene 1, a moralizing trio castigating bribery at the end of Part I, and a poignant dramatic trio at the end of Part II. As the sentiments of the characters involved in the

Ex. 10.5

Ch'ab-bia pa - ce il mio cor non è pos - si - - bi-le, non è pos - si - - - - bi - le

(It is not possible that my heart should be at peace)

ensembles are virtually the same, Gianettini's settings are euphonic, with only occasional flights of imitative counterpoint. The opening phrases of the duet 'Sgombreranno un di' typify his relaxed, Venetian style of counterpoint (Ex. 10.6).

Ex. 10.6

(One day they will clear from my heart, cruel destiny ‖ One day they will clear from my heart, grievous cares)

In the final trio, Moses and Zipporah sing together in flowing phrases of parallel thirds and sixths, until the temptation to set 'Ti segue costante' as a little fugato persuades Gianettini to separate them for a second or two. Jethro, in unison with the basso continuo for most of the trio, is allowed the last 'Addio' on a tierce de Picardie as he rides off into the sunset.

11

Palermino's *Il Sansone* (1688)

Il Sansone, an oratorio in two parts
Music by Bonaventura Aleotti *detto* Padre Palermino
Libretto—anon.

Characters:		
	Dalida (Delilah)	S.
	Inganno (Deceit)	S.
	Filisteo Capitano (Captain of the Philistines)	A.
	Sansone (Samson)	T.
	Morfeo (Morpheus)	B.
	Chorus *a 5*	S. S. A. T. B.

Instrumental accompaniment *a 4*
First performed in Naples in 1686; revived in San Carlo rotondo in 1688.

In all Modenese sources, the composer of *Sansone* is identified as Padre Palermino. As this name suggests, he was a native of Palermo in Sicily, but spent the middle years of his career in Ferrara, the one-time capital of the Este dynasty. The first of his oratorios to be produced in Modena was *Il trionfo della morte* (1685). Another, *S. Rosalia*, was performed with *Sansone* in the season of 1688.

The Libretto

This 1688 version of *Sansone* treats the same events from the Book of Judges as Giardini's oratorio of 1680 (see Chapter 3). It begins with the triumph of Samson over the Philistines and ends with his capture and humiliation. But the dramatic treatment of the story is markedly different. The anonymous librettist shows little interest in moral questions, or in tracing the psychological tensions aroused by Samson's passion for Delilah, or in heightening the tragedy of the hero's downfall, as Giardini does so effectively; his only concern is to show how a man of valour can be destroyed by an unholy alliance of human and spiritual enemies. The chief enemy to be confronted is flattery, as the disconsolate Samson informs the audience at the end of the oratorio, 'only he who can resist flattery will receive garlands of glory'.

In casting the story in dramatic form, the librettist dispenses with a narrator. Two allegorical characters, Deceit and Morpheus, join

Delilah in her scheming, and a Chorus of Philistines is given a prominent role. The dramatic structure of the oratorio is as follows:

Sinfonia

Part I:

Scene 1   Chorus, Samson, Captain: The Philistines rejoice at hav-
             ing captured Samson. He mocks them, breaks his chains,
             and puts them to flight. The Captain vainly strives
             to rally his men. He prays to the stars for the ruin of
             Samson and the fall of Israel.

Scene 2   Delilah, Captain: Delilah overhears him, and echoes his
             prayer. The Captain persuades her to use her beauty to
             ensnare Samson.

Scene 3   Delilah, Deceit: Delilah calls Deceit to her aid. He advises
             her to flatter Samson and confuse him by mixing tears
             and smiles.

Part II:

Scene 4   Samson, Delilah: Samson discovers Delilah in great dis-
             tress, weeping. He comforts her and asks her to stay with
             him. They make a pledge of fidelity for life.

Scene 5   Deceit, Morpheus: Deceit echoes the pledge, approv-
             ingly. He invokes Morpheus in Erebus, asking him
             to drug Samson to sleep so that Delilah can make her
             assault. Morpheus weaves his spell.

Scene 6   Delilah, Deceit, Samson: The temptation. Delilah,
             prompted by Deceit, accuses Samson of not loving her
             and of treating her wishes lightly. She wants to know the
             source of his great strength, to which he replies, 'The
             strength of Samson is knotted in his hair.' As Delilah
             rejoices, Samson is overcome with fatigue. She lulls him
             to sleep.

Scene 7   All: The catastrophe. Morpheus and Deceit join Delilah
             to charm the sleeping hero. She cuts off his hair, then
             rouses him to face his enemies. The Philistines attack
             him, disarm him, and bind him. The Captain holds up
             his shorn locks as a trophy. Samson, echoed by the
             Chorus, warns the audience to resist flattery.

The libretto contains very little recitative; the dialogues in Scenes 2
and 4 are the only places where the norm is approached. For the rest
of the drama, arias and ensembles predominate. They vary in metrical
structure from the tetrasyllabic lines of the Philistine army in flight:

> Fugga chi può.
> Rugge Sansone

Il fier Leone
Si scatenò.
Fugga chi può.

(Fly if you can. Samson roars. The fierce lion has broken loose. Fly if you can.)

to the hendecasyllables of Delilah's aria, sung while she is stroking her compliant lover's hair:

O crini, ò bei legami,
O degli affetti miei dolci catene,
O di mia vita preziosi stami
Intessete ghirlande alla mia speme.

(O hair, O blessed bonds, O sweet chains of love, O precious threads of my life, weave garlands for my hopes.)

This aria begins with an echo of Samson's last phrase in the foregoing recitative, 'al crine'. Verbal echoes of this kind are habitually used by the librettist to make connections from one character to another, or between one scene and the next. At the end of Scene 1, for example, the Captain, after praying to the stars for help predicts:

Se rovina Sanson cadrà Israelle.

(If Samson is ruined, Israel will fall.)

Delilah enters (Scene 2), pondering what the Captain has just said, and decides on a course of action (in recitative):

Se rovina Sanson, cadrà Israelle.
Spirti audaci, all 'impresa.

(If Samson is ruined, Isreal will fall. Bold spirits, go to work.)

Such echo devices have a long pedigree in dramatic literature, stretching back to antiquity. In the sixteenth century the echo was commonly used in pastoral plays and in the spiritual dialogues of the Oratorians. It survived well into the seventeenth century as a picturesque feature of oratorio librettos but was a rarity by 1680. It is surprising to see this relic of a bygone age crop up so frequently in *Sansone*.

Allegorical characters appear in all Palermino's extant oratorios. The pair in *Sansone*, Inganno and Morfeo, regularly took the stage in the opera house as well as in the oratory. In Palermino's oratorio they function as agents of Delilah. Deceit is her closest ally; she hails him in Scene 3 as her inseparable companion—'compagno mio indivisibile'— and uses him as a go-between to enlist the occult powers of Morpheus. Their presence alongside Delilah at the catastrophe ensures that the final scene of the oratorio is replete with ensembles.

The Score

Sansone is scored for five solo voices. Duets and trios are clearly labelled with the names of the characters who sing them, and the choruses, labelled simply '*a 4*' and '*a 5*', can all be performed by the combined voices of the soloists. The accompanying strings *a 4* play the Sinfonia and are involved in all but five of the twenty-two formal numbers. The distribution of formal numbers among the five soloists is shown in Table 11.1.

TABLE 11.1. Distribution of Numbers in *Sansone*

	Delilah S.	Deceit S.	Captain A.	Samson T.	Morpheus B.
Continuo arias (2)	1	0	0	1	0
Arias with ritornellos (2)	1	0	0	0	1
Accompanied arias (8)	4	2	1	1	0
Continuo duet (1)	1	0	0	1·	0
Duets with ritornellos (3)	2	1	1	1	1
Continuo trio (1)	1	1	0	0	1
Trio with ritornello (1)	1	1	0	0	1
Chorus with continuo (1)	1	1	1	1	1
Accompanied choruses (3)	3	3	3	2	3

There are two misleading copyist's errors in the Modenese score (*I-MOe* Mus. F. 886): the lowest vocal part in the opening Chorus carries a tenor clef when it should be bass, and, more seriously, the notes in the two upper string parts of Delilah's aria 'Molle sonno' (Scene 6), make musical sense only with soprano clefs, not treble as copied. Palermino evidently intended the parts for violettas.

The oratorio is introduced by a festive sinfonia in D major. Three musical ideas, a solemn Adagio (A), a gigue-like Allegro (B), and a pompous processional employing echo effects (C), are arranged in a pattern that relies for its stability upon a balanced tonal scheme: AB (in D major), AB (in A major), C (in B minor and E minor), BC (in D major). At the end of this sequence, Palermino changes the metre to 6/8 for what appears to be a lively fourteen-bar coda. As the Philistines burst into song, however, using the same motifs in 6/8, we realize that the last section of the Sinfonia is indeed a ritornello for the chorus 'Speranze, gioite' (Ex. 11.1). The flow of contrapuntal invention for chorus and orchestra continues for 140 bars, with only brief

Ex. 11.1

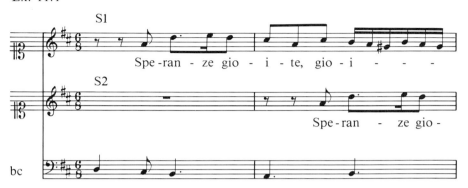

(Rejoice in hopes)

pauses for Samson's scornful interjections in arioso recitative. The first eighteen double-pages of the score show that Palermino designs his music on a grand scale, using sectional repeats and a carefully devised tonal scheme to consolidate his ideas.

It is interesting to note that Samson's interjections, though intended by the librettist to be parodistic echoes of the Chorus's jubilation, are not set as musical echoes. Here, as elsewhere in the oratorio when faced with poetical ingenuity, Palermino ignores it. His interests are not literary, but musical. His method of composition is to search the text for some phrase which stimulates a musical idea capable of extensive development. Having conceived the idea, he then explores its possibilities until his powers of invention and acute sense of formal balance are satisfied. This single-minded approach produces splendid

music, but may not always fulfil the librettist's good intentions. A
case in point is Samson's aria in Scene 1, 'Cingetemi pur le piante'.

The very first word of the text—'Cingetemi' ('Bind me')—suggests
to Palermino a ground bass. Moreover, because Samson describes his
chains as 'double' and 'heavy', the ground bass is short (two bars
long) and energetic, making the voice above it struggle hard for
freedom (Ex. 11.2). Occasional half-bar rests between statements of
the ostinato are perhaps meant to suggest that the chains are break-
able. The last line of the four-line lyric depicts Samson throwing his
broken chains to the ground: 'Mirate per terra infrante'. Palermino,
however, is so absorbed in developing his ground bass that he ignores
the dramatic implications of the text until he reaches the coda, where
rhetorical cries of 'mirate, mirate' raise expectations that the ostinato

Ex. 11.2

Ex. 11.2 *(cont.)*

for - - - - - - - - - - - -

10

- - - - - - - te con

dop-pia ca-te- na e for - te

(You still bind me, ropes, with a strong double-chain)

will lose its grip. It doesn't, for purely musical reasons. In Palermino's grand design, the ground bass is heard twenty times: six times in B minor, once in D major, seven times in F sharp minor, twice in A major, and four times in B minor.

Delilah's accompanied arias in Scene 6 demonstrate Palermino's schematic but adept handling of the orchestra. In 'Ò crini, ò bei legami', Delilah is caressing Samson's hair (Ex. 11.3). Her ecstatic song is accompanied chiefly by the continuo, while the strings are used to echo or anticipate her musical phrases, as though giving her a breathing space to fondle his locks. In 'Molle sonno', where she is charming the hero to sleep (Ex. 11.4), Palermino, having replaced the violins with violettas, allows the strings *a 4* to cover the whole of her incantation in an unbroken veil of sound. At the end of this through-composed aria the orchestra restates, in a brief da capo, the gloomy notes of the opening phrase.

Ex. 11.3

(O hair, O blessed bonds)

Ex. 11.4

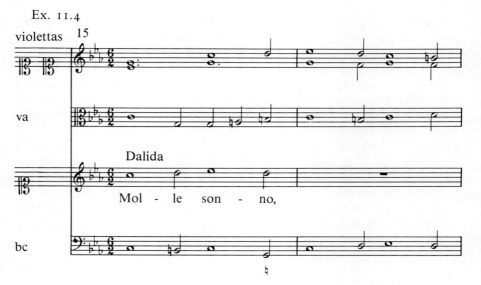

Ex. 11.4 *(cont.)*

son - - no deh stil - - la_____

(Soft sleep, ah! drop)

As the last notes of the coda die away, Delilah is joined by Deceit
and Morpheus for a trio of enchantment, which begins unaccompanied (Ex. 11.5). The textures of 'Chiusi i lumi' vary from three-part
counterpoint to brief solo passages. At its climax, their conspiratorial
creed, 'Shade is an opportunity for treachery', is sung homophonically,
and then contrapuntally before the strings conclude with a seven-bar
ritornello. For dramatic reasons, this trio is more loosely constructed
than other ensembles in the oratorio, but its varied textures are held
together by a logical tonal plan: F major, B flat major, G minor, E flat
major, F major.

Ex. 11.5

Dalida

Chiu - se i lu - mi il fier ti- -ran - no, il

Inganno

Chiu-se i lu - mi il

Morfeo

Chiu-se i lu - mi - il fier ti -

Ex. 11.5 *(cont.)*

(The fierce tyrant closes his eyes, and to Deceit . . .)

Palermino's contrapuntal dexterity is seen at its best in the second chorus in Scene 1, depicting the panic and flight of the Philistines. The fugal style is called for, and Palermino writes a fugal quintet on three subjects, combining the first with the second and third in the following manner:

A   Fugga chi può (for S. S. A. T. B.)   (Ex. 11.6)
B   Rugge Sansone (for A. T. B.) combined with A (for S. S.)
C   Si scatenò (for A. T. B.) combined with A (for S. S.)
A   (for S. S. A. T. B.)
B   (for S. S. A.)
C   (for S. S. A.)
A   (for S. S. A. T. B.)

By the end of the quintet, all the voices have had a share in the development of the three fugue subjects.

Ex. 11.6

(Fly, fly if you can)

The most elaborate numbers in the oratorio, the accompanied choruses, occur in its opening and closing scenes. Palermino relished a technical challenge, and took delight in composing vocal and instrumental textures in as many as seven real parts. For technical assurance, and skill in organizing large-scale musical forms, Palermino deserves recognition as one of the most accomplished composers of the late seventeenth century. More's the pity that so little of his music is extant.

12

Vitali's *Il Giona* (1689)

Il Giona, an oratorio in two parts
Music by Giovanni Battista Vitali
Libretto by Abbate Domenico Bartoli
Characters: Dio (God)   B.
              Angelo (Angel)   S.
              Giona (Jonah)   A.
              Capitano (Captain)   T.
              Servo (Servant)   S.
              Rè (King of Nineveh)   B.
              Regina (Queen of Nineveh)   S.
              Chorus *a* 5   S. S. A. T. B.
Instrumental accompaniment *a* 4
First performed in the oratory of San Carlo rotondo, Modena, in 1689.

Giovanni Battista Vitali (b. 1632, Bologna; d. 1692, Modena) had
been sottomaestro at the ducal court in Modena since the restoration
of the cappella in 1674. Before coming to Modena, he had composed
oratorios in Bologna—*Agare* (1671) are *Gefte* (1672)—and a *Sinfonia di
vari strumenti* contributed to Pratichista's oratorio, *Il trionfo della fede*
(1672). His duties at Modena, however, were primarily in the instrumental field. It was not until the hasty departure of Maestro de
Grandis from the court in 1683 had brought additional responsibilities
to the sottomaestro, that Vitali returned to vocal composition, producing cantatas for the Accademia de' Dissonanti, hymns for the ducal
chapel, and two oratorios for San Carlo rotondo, *L'ambitione debellata*
(1686) and *Giona* (1689). Abbate Bartoli was a native of Lucca. *Giona*
appears to have been his sole contribution to the oratorio repertory.

The Libretto

Bartoli was a librettist of the old school, paying scant regard to
Spagna's notions about oratorio becoming 'perfect spiritual
melodrama'. The action in Bartoli's oratorio, based on the Book of
Jonah, ranges over a time-span of several weeks and visits various far-
flung locations: Heaven, Joppa, the Mediterranean Sea, and the royal

palace in Nineveh. There are solo roles for seven characters, including a major role for God, and if we add the variety of roles played by the Chorus (i.e. avenging angels, moral pedagogues, narrator, ship's company, and Ninevite courtiers) we reach a grand total of twelve. Only two other oratorios in the Modenese repertory compare with it: Pallavicino's *Il trionfo della castità* (1685) with twelve characters, and Palermino's *Il trionfo della morte* (1685) with eleven, including four roles for the Chorus. Bartoli's one concession to modern taste is to eliminate the role of Narrator; a story as familiar as Jonah's hardly required one.

The dramatic structure of the oratorio is as follows:

Sinfonia
Part I:
Scene 1   Chorus, God, Angel: The Chorus of avenging angels urges God to strike Nineveh with thunderbolts. He gives them leave to 'Go, burn, and destroy' the city. The guardian Angel of Nineveh intervenes and persuades God, out of fatherly love, to relent. The Chorus reflects on God's mercy.

Scene 2   God, Jonah, Angel: God calls Jonah, and sends him to warn the Ninevites. Jonah hesitates. The Angel advises obedience, but fear prompts Jonah to attempt an escape by sea.

Scene 3   Chorus, Captain, Jonah: The Chorus, acting as narrator, describes a lovely dawn. The Captain and his crew prepare to sail. Jonah is welcomed aboard.

Scene 4   God, Captain, Jonah, Chorus: God sees Jonah's cowardice and conjures up a storm. The panic-stricken Captain agrees to Jonah's request to be thrown overboard. The Chorus describes the sea's return to calm.

Part II:
Scene 5   Servant, King, Queen, Chorus: At the court of Nineveh a Servant announces to the King and Queen the arrival of Jonah in the city, predicting that the end is nigh. The King sends for Jonah. The Chorus reflects on the sinfulness of potentates.

Scene 6   Jonah, King, Queen, Chorus, God: Jonah appears and presents his credentials (his story). His mission to be a scourge to Nineveh horrifies the King and Queen. Jonah calls upon them to repent. The whole court capitulates and cries for mercy. The King throws away his sceptre, the Queen abandons her jewels. God relents and pardons them. The royal couple rejoice. The Chorus calls upon

mortal men to live the kind of lives that break the sternness of the Eternal One.

The key to Bartoli's presentation of the story is the King of
Nineveh's command to his erring subjects in Jonah 3: 8–9: 'Yea, let
them turn every one from his evil way. . . . Who can tell if God will
turn and repent, and turn away from his fierce anger, that we perish
not?' The fierce anger attributed to God inspired the librettist to
open his oratorio with a celestial scene in which God himself, after
fulminating like a stage Lucifer, is persuaded to repent by the Angel.
This curious dramatic concoction of God–Lucifer, who provokes and
dominates the action, is required to repent a second time in Scene 6,
when the Queen, having jettisoned her rich baubles as a sign of her
own contrition, demands a quid pro quo:

QUEEN: Odi i miei prieghi e poi,
                  A miei sopir nega pietà, se puoi.
GOD: Non più, non più singulti.
                  Intenerito io sono.
                  Monarchi afflitti asserenate il ciglio.
                  Mutaste il cor; io pur mutai consiglio.
                  Voi perdon mi chiedeste, io vi perdona.

(Hear my prayers and then, for my relief deny pity if you can. ‖ No more,
no more sobbing. I have softened. Distressed monarchs, dry your eyes. You
changed your hearts; I, too, have changed my decision. You asked me for
pardon, I pardon you.)

By wrenching a primitive concept of God from its proper context in
the Old Testament and presenting it as a living character on the stage
(albeit the stage of an oratory), Bartoli can be seen to be violating his
source for dramatic effect. Credulity is strained to its limits as this
theatrical God raises a storm in Scene 4:

                  Sù, sù, si conturbi l'etra.
                  Si scatenino i venti,
                  Sorga fiera tempesta.
                  E finche il rio fellon non cada al fondo,
                  Si sconvolgan le nubi, il mare, il mondo.

(Up, up! Stir yourself, air. Winds break loose, fierce tempests blow! And until
the wicked traitor falls into the depths, overturn the clouds, the sea, the world.)

If one passes over these extraordinary lapses of theological decorum,
to consider only the dramatic quality of the libretto, *Giona* is quite an
effective piece. Part I has variety enough for any taste, and Part II,
with the whole of Nineveh wailing in sackcloth and ashes, sustains the
penitential mood impressively.

On the title-page of the Modenese score of *Giona* (*I-MOe* Mus. F. 1260), it is described as an 'oratorio a 5 voci con stromenti'. An unknown hand has later corrected the 'a 5' to 'a 7', which accords with the full list of solo roles. The doubts raised by this tiny amendment persist, for, in truth, the score has plenty of work for as many as twelve singers. It can be performed by six, however, if the second soprano takes the role of Angel in Part I, and Servant in Part II.

The orchestral music is straightforward, being scored for a string ensemble of two violins, viola, and basso continuo. This standard arrangement is changed only once, for the Captain's aria 'Chiaro il Ciel' in Scene 3. Here, the viola is left out in order to highlight a remarkable canonic ostinato played by the two violins, against which the Captain (singing in the viola register) and the basso continuo weave their free contrapuntal lines. Though the strings are not required to accompany duets or choruses, they participate in fifteen of the sixteen arias, and open the proceedings with the customary sinfonia.

Vitali was 57 years old when he composed *Giona*. Setting to music an old-fashioned libretto containing a large number of choruses and duets must have reminded him strongly of his Bologna days. This new task posed no fresh challenges; indeed, it offered him an amenable show-case for his artistry.

The oratorio begins with a sinfonia in B flat major (Ex. 12.1). The solemnity of the first section, with its chains of 2–3 suspensions, is dispersed by the jaunty pomposity (alla Francese) of the second. At the end of the Sinfonia the Chorus makes an arresting entry, singing a brilliant toccata *a 5*, and hurling down its thunderbolts with gusto

Ex. 12.1

Ex. 12.1 (*cont.*)

(Ex. 12.2). In madrigalean manner, 'divine tolerance' is given a
grand homophonic setting in G minor, while its transformation 'into

Ex. 12.2

(Let fly now your arrows)

'rigidity' is symbolized by a strict fugato. The scale and complexity of this chorus is a broad hint that the oratorio is going to prove that the 'moderns' who disdain the old craft of vocal polyphony are misguided. By the end of the tenth chorus in *Giona*, even the most sceptical of the moderns must have been impressed by Vitali's demonstration of old skills.

The choruses are the most elaborate set pieces in the oratorio. One can see from the distribution of arias and ensembles among the singers in Table 12.1, that the soloists must have spent much of their scheduled rehearsal time working on their intricate textures.

TABLE 12.1. Distribution of Numbers in *Giona*

	Queen S.	Angel S.[a]	Jonah A.	Captain T.	God B.	King B.
Continuo aria (1)	0	0	0	1	0	0
Arias with introd. (5)	1	2	0	0	1	1
Arias with coda (3)	0	0	1	0	2	0
Accompanied arias (7)	1	0	1	2	2	1
Duets (4)	4	0	0	0	0	4
Choruses (10)	10	10	10	10	4[b]	6[b]

[a] The second soprano, when taking the part of Servant, sings only recitative.
[b] Assuming the choral bass part is sung by the King in Part I, and by God in Part II.

The four duets are sung by the King and Queen. As their thoughts and feelings are identical, Vitali sets their matching phrases in imitative counterpoint with continuo support, for much of the time in unison with the King (Ex. 12.3).

Ex. 12.3

Ex. 12.3 *(cont.)*

(The archer's bow of divine correction, a soul who is sad . . .)

Vitali's arias are expressive and shapely, but remarkably brief for the late 1680s; few exceed twenty-four bars in length. A typical aria is the King's 'Il pentirsi' in Scene 6 (Ex. 12.4). In the solemn introduction for strings, the opening motif, incorporating a striking diminished fourth, is treated canonically. A more extended canon between the continuo (dux) and the bass voice establishes a scholarly, contrapuntal manner which persists to the end. The form of this aria, ABB, is much more frequently used by Vitali than the fashionable ABA form.

Ex. 12.4

Ex. 12.4 *(cont.)*

gan - - - - - no che di Di - o

lo sde-gno_at-ter - - - [- ra]

(Penitence is a sweet trick to demolish God's anger) (If hitherto I was all immersed in delights too tender.)

The climax of the oratorio occurs in Scene 6, when God is moved by the plaints of the Queen to alter his stern decree. For this crucial situation, Vitali composes for the Queen a beautiful accompanied aria of exceptional length (seventy-six bars), 'Frà delitie troppo tenere' (Ex. 12.5). Following a fugal introduction of fourteen bars, the Queen sings a two-bar 'motto' based on its opening, to which the orchestra responds with a suave four-bar parody. The preliminaries over, the motto is restated and extended in rising sequences and flowing *passaggi* on 'immersa'. Again the orchestra parodies. The music moves into minor keys as her remorse deepens. A realistic sob on 'palor di mesta ce-nere' colours the text. At the end of the aria the Queen reflects that it is only the wasted life that perishes in the fires of death. This moral

Ex. 12.5

strs.

bc

Ex. 12.5 *(cont.)*

conviction is stated, first with a continuo accompaniment in G minor, and then with full orchestra in F minor. A tierce de Picardie restores the major mode at the final chord. In this through-composed aria, Vitali achieves a sense of unity, not by statement and restatement of main themes, but by maintaining an eloquent dialogue between the soloist and the orchestra. At times the orchestra anticipates her thoughts, at others it elaborates upon them in easy-flowing counterpoint. Only in the final phrase do they join forces in rich, five-part harmony.

At the end of Chapter 3, I remarked on the conservatism of Ferrari's setting of *Sansone*. Vitali's *Giona*, performed in Modena nine seasons later, is also conservative in form and style. Whether the courtiers found the music to their taste or not, Vitali's score gave them the chance to sample musical craftsmanship of the highest level.

13

Pistocchi's
Il martirio di S. Adriano (1692)

Il martirio di S. Adriano, an oratorio in two parts
Music by Francesco Antonio Pistocchi
Libretto by Silvio Stampiglia
Characters: Adriano   A.
                    Natalia (his Christian wife)   S.
                    Claudio (a minister at court)   T.
                    Massimiano Imperatore (Emperor) B.
Instrumental accompaniment *a* 5
First performed in Modena in 1692.

In the early 1690s Pistocchi (b. 1659, Palermo; d. 1726, Bologna) was a famous professional singer in the opera-houses of northern Italy. His chief patron, Ranuccio Farnese II, duke of Parma, was intimately acquainted with the Estensi, having married in succession two of Francesco II's aunts. In 1692 the compliment was returned when Francesco took as his bride his first cousin, Princess Margherita Farnese. The celebrations, which took place in July and August, included a performance of Gianettini's opera *L'ingresso alla gioventù di Claudio Nerone*, at the Fontanelli Theatre in Modena in which Pistocchi, as the Duke of Parma's most celebrated 'musico', sang on the stage alongside his counterpart, the duke of Modena's Siface. Though the libretto of *S. Adriano* contains no details of the circumstances of the performance in 1692, its subject-matter, matrimonial fidelity, was a most appropriate topic for gracing a royal wedding. As the title-role is cast for an alto voice, Pistocchi himself may have sung the part at the première. Four years later, when the oratorio was revived in Modena, Luigi Albarelli sang the alto part; the other members of the cast were Marc Antonio Origoni (S.), Corregio of Mantua (T.), and Antonio Pietrogalli (B.).

The Roman poet Silvio Stampiglia (1644–1725) is best known for his collaborations with Giovanni Bononcini and Alessandro Scarlatti. He was a member of Queen Christina's circle and in 1690 became a founder-member of the Accademia dell' Arcadia, a group of poets, musicians and artists devoted to the study of classical culture. His oratorio *S. Adriano* is set in Nicomedia, Asia Minor, a province of imperial Rome.

The Libretto

According to legend, Adriano was a Roman official at Nicomedia at the end of the third century AD. After observing the endurance of Christians under torture, he embraced their faith and was thrown into prison. His young Christian wife Natalia visited him there to support him with her prayers. She witnessed his torture and execution. In Stampiglia's dramatization, the events are placed in the reign of the Emperor Maximian, who confronts and punishes Adriano for apostasy. Adriano's former friend Claudio acts for the emperor in trying to persuade the saint to come to his senses, but to no avail.

In constructing his oratorio, Stampiglia followed the precepts of Arcangelo Spagna as closely as the librettist of *S. Teodosia* (see Chapter 7). The two oratorios are indeed quite similar in design, having many short scenes shared by the four main characters. The dramatic structure of *S. Adriano* is as follows:

Sinfonia
Part I:

Scene 1	Adriano, Natalia, Claudio: Adriano is comforted by Natalia as he weeps over the fate of the Christian martyrs. He declares himself a convert. Claudio visits their home to warn Adriano of the danger of defying the emperor. Adriano is willing to die for his new faith.
Scene 2	Massimiano, Adriano: The emperor examines Adriano, and sends him to prison.
Scene 3	Claudio, Adriano: Claudio conducts Adriano to his cell. The saint is unrepentant, welcoming hardship.
Scene 4	Claudio, Massimiano: Claudio reports back to the emperor who, in a fury, condemns Adriano to death. Claudio pleads for mercy, and time for Adriano to change his mind. Massimiano wavers.
Scene 5	Claudio, Natalia, Adriano: Claudio directs Natalia to Adriano's cell. She weeps over his chains, and comforts him.

Sinfonia
Part II:

Scene 6	Claudio, Adriano: Claudio challenges Adriano's blind faith. The saint ponders the mystery of the Holy Trinity

and rejects his friend's false reasoning. As Claudio
leaves, Adriano sings of the dying swan.

Scene 7   Claudio, Massimiano: Claudio reports to the emperor
that Adriano mocks his power. Massimiano reiterates the
death sentence and invokes the spirits of Hell.

Scene 8   Natalia: Natalia prays to Heaven, enraptured by God's
goodness.

Scene 9   Claudio, Adriano: Claudio brings news of the death
penalty. He gives the saint freedom to pay a last visit
to his wife.

Scene 10   Natalia, Adriano: Natalia is shocked at seeing her husband released, and home again. She refuses to admit
him, accusing him of lying and loss of faith. When he
tells her of his sentence, she relents and embraces him
for the last time. Adriano returns into the custody of
Claudio.

Scene 11   All: Natalia, Claudio, and Massimiano witness Adriano's
torture and martyrdom.

The story unfolds without recourse to formal narration; each
character is named on entry, or before. Natalia's first words in Scene 1
are 'Adriano, my husband'. His reply, 'Natalia, I tell you the truth',
identifies her. Claudio's entry is prefaced by Natalia's 'But Claudio is
coming to see us'. By similar promptings, the audience gathers from
the dialogue where events are taking place: Adriano in Scene 10 cries,
'Darling, why are you closing the door of our dear home on me?' The
action flows smoothly between Adriano's home, his cell, the imperial
throne-room, and the place of execution. With each change of location, Stampiglia is careful to allow time to elapse by inserting a
reflective aria or a brief soliloquy. Adriano's aria about the dying swan
at the end of Scene 6 gives time enough after Claudio's exit from the
cell for him to reach the throne-room to report to the emperor in
Scene 7. Seven arias are used in this way, to mark the passage of time
when a change of location is impending.

The subtlety of Stampiglia's stagecraft is matched by subtlety of
characterization. Each of the four principals is a rounded personality
drawn, as it were, from life. Natalia, for instance, might have fitted
the stereotype of a peerless and faithful wife had she not under-
estimated her husband's honour and barred the door to him. From a
woman of tender spirituality who sings in Scene 8, 'Dear God, how
gentle you are', one hardly expects the torrent of abuse that pours
from the window of her house in Scene 10: 'wicked husband', 'liar',
'fool', and 'idolator'. Adriano's news of the death penalty calms her
down, restoring her faith in him:

NATALIA: Và pur lieto a morire.
Mà pria frà queste braccia
Gl'ultimi amplessi miei ti prendi e godi.
(Go now happy to your death. But first, in these arms, take and enjoy my last embrace.)

The spirit of Arcadia permeates much of the poetry in *S. Adriano*. The saint likens himself to the dying swan and the mythical pheonix, and Natalia prays to the Lord of benevolent nature to bless her husband. At the end of Scene 9, as Adriano makes the journey home, Claudio passes the time singing a madrigal of the spring, though one with a sharp reminder that nature is not always as benevolent as some Arcadians might imagine:

Quando ride ameno il Maggio
Di bei fior si smalta il prato;
Gode il mirto e gode il faggio
L'armonie di stuolo alato.
Rozzo monte è belva fiera
Al fulgor di primavera
Di più gioie ancor s'adorna.
E pur se nasce il verno April ritorna.

(When pleasant May is smiling, the meadow is adorned with pretty flowers; myrtle and beech rejoice in the harmony of the birds. A craggy mountain is a wild beast in the splendour of spring, decked with many precious stones. Only if winter is born can April return.)

Stampiglia's main purpose in *S. Adriano*, however, is to examine the conflict between Christian and pagan values. The moral arguments are found chiefly in the exchanges between Adriano and Claudio. Their first disagreement is over the value of human life. 'Do you care nothing', asks Claudio, 'about losing your life and obscuring your fame?' Adriano replies, 'I intend to lose my life and acquire fame.' After being confined to prison, Adriano mounts an attack upon the behaviour of emperors: 'Wicked kings sometimes reward lions, yet punish the righteous.' Claudio responds obliquely with, 'Is then the emperor a tyrant?' To this dangerous question, Adriano's reply is unflinching: 'He is an idolator, and that is enough.' The dispute enters its final phase in Scene 6. Claudio accuses Adriano of having a blind faith in a god he cannot understand. The saint responds with a brief meditation on the mystery of the Holy Trinity—'three distinct lights shining as one'—but, after admitting his own incomprehension, asserts that, nevertheless, his blind eyes will one day open to joy, while Claudio's will open to grief. There they rest their cases, the argument unresolved. At the end of the oratorio, pagan Claudio

celebrates the death of a traitor to the State, Christian Adriano bathes
in the sweetness of martyrdom.

The Score

The oratorio is scored for four solo voices. No choruses are called for
in the libretto, and there are only two ensembles: a duet for Adriano
and Natalia at the end of Part I, and a shared duet for all the principals
at the end of Part II.

The accompaniment is for a string ensemble *a 5*: two violins, alto
viola, tenor viola, and basso continuo. Pistocchi uses the full ensemble
for the sinfonias and duets, and for the majority of arias. In some
arias, however, the scoring is reduced or modified to suit the situation. The alto viola is omitted in Adriano's aria di bravura 'La speme
mi dice' to avoid needless competition with the voice, the bright tone
of the violins is excluded from Massimiano's sombre aria 'Più non
m'ingombri il petto', and two solo cellos, playing in high register
(tenor clefs), are the elegant companions of Adriano as he sings of the
dying swan.

There is some evidence in the score that Pistocchi planned to use a
large string ensemble with, perhaps, three or four players to each part.
In the middle section of the first Sinfonia, the violin staves carry
markings of 'soli' and 'tutti', suggestive of a concertino/concerto
grosso division of forces. In addition, all the string parts accompanying Adriano in the final duet of Part II carry the rubric 'concertino
piano', a clear instruction to play the section quietly with only one
instrument to a part. These hints may not be sufficient to prove that a
large orchestra was used, but they show, together with the careful
scoring of the rest of the oratorio, that Pistocchi was an expert
colourist.

The two sinfonias in *S. Adriano* are very tuneful. The first begins in
cheerful mood, with upper and lower strings engaging in playful
imitation (Ex. 13.1*a*). A transitional section of tender pathos (Ex.
13.1*b*) leads to gently-flowing dance measures in 3/4, in which two
solo violins take the lead (Ex. 13.1*c*) and the 'tutti' responds. A repeat
of the opening section rounds off the Sinfonia, producing the composite design:

A	Allegro in F major	15 bars
B	Adagio in D minor	7 bars
C	Binary-form dance in D minor	32 bars
A	Allegro in F major	15 bars

Ex. 13.1

Ex. 13.1 *(cont.)*

The second sinfonia (before Part II) is a majestic processional in the French style. It is in binary form, with each section repeated; there are even alternative phrases written on the bass stave for first-time and second-time performance of the sections. This brief sinfonia is on page 42^r of the manuscript. As pages 42^v–4^v are left blank, it looks as though Pistocchi intended to add further sections to the Sinfonia but, for some reason, never got round to composing them.

    The arias and duets are distributed among the characters as shown in Table 13.1. If one adds to this list Adriano's accompanied recitative in Scene 6, it is clear that Pistocchi regarded the orchestra as á vital component in the development of his musical ideas. It offered him variety of colouring, opportunities for delicate interplay with the

TABLE 13.1. Distribution of Numbers in *S. Adriano*

	Natalia S.	Adriano A.	Claudio T.	Massimiano B.
Continuo aria (1)	0	0	1	0
Arias with ritornellos (7)	3	3	1	0
Accompanied arias (9)	1	2	3	3
Accompaned duets (2)	1.5	1.5	0.5	0.5

voices, and the chance to expand his formal designs by means of long ritornellos and interludes.

Natalia's aria 'Mio Signor' (Ex. 13.2), is a prayer that her newly converted husband will be blessed by a ray of light from heaven. The musical ideas that are to sustain her prayer are presented in the opening bars of the orchestral introduction. The tonic pedal-point signifies, perhaps, God's reliability. The sarabande motif given to the second violin later carries her address to God, 'Mio Signor', just as the first violin's response carries the essence of her prayer, 'Vibra un raggio' ('send a ray of light'). Though Pistocchi uses a solo cello (tenor clef) in the accompaniment, it is not for the purpose of solo display, but to add intensity to the discordant harmony above the pedal. By using the orchestral introduction as an interlude and coda, and setting the text in

Ex. 13.2

Ex. 13.2 (*cont.*)

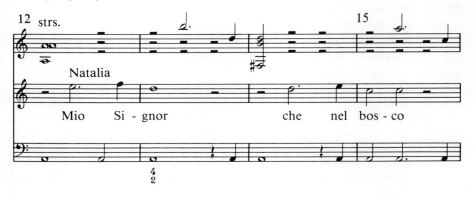

(My Lord, who knows how to defend in the wood and meadow . . .)

a through-composed manner, he produces the following composite design:

A	Orchestral introduction in A minor	12 bars
A	'Mio Signor' in A minor	12 bars
B	'Le bell'erbe' in A minor	10 bars
A	Orchestral interlude	12 bars
C	'Tu dai cieli vibra un raggio' in E minor	15 bars
C	'Tu dai cieli' repeated in A minor	15 bars
A	Orchestral coda	12 bars

This aria is of ample proportions (eighty-eight bars), but seems quite modest when compared to the mammoth da capo structure Pistocchi devises for the end of Part I, the duet of Adriano and Natalia in prison (Ex. 13.3):

Ex. 13.3

(Darling, darling, if you truly love me those harsh bonds of mine . . . ‖
Mistress, mistress, if you truly love me these harsh bonds of mine . . .)

A	Orchestral introduction in G major	40 bars
B	Duet, 'Caro/Bella se ver che m'ami' in G major–A minor–D major–G major	52 bars
A	Orchestral interlude	40 bars
C	Solo exchanges accompanied by two tenor violas in E minor–B minor–E minor	30 bars
A	Orchestral interlude	40 bars
B	Duet, 'Caro/Bella'	52 bars
A	Orchestral coda	40 bars

When one reflects that the 262 bars of this symmetrical structure are
generated by a mere three lines of verse in which a husband and
wife pledge their love in affliction, one wonders whether Pistocchi's

enthusiasm to produce an impressive finale has got the better of his dramatic judgement. The double duet 'Languisci e mori' at the end of Part II is another symmetrical structure, satisfying from a musical point of view, but dramatically inappropriate. Claudio and Massimiano relish the death of Adriano (section A); Natalia and Adriano affirm their faith in God, and Adriano's soul flies to heaven to the sound of an orchestral flourish (section B); and then the heathen pair return to sing an exact reprise of A, quite unaffected by, if not ignorant of, the saint's passing.

Pistocchi was a composer of great lyric charm and, as we have seen, a subtle colourist, but he had little talent for effective dramatic writing. His settings of recitative follow the natural rhythms and inflections of speech, as in Adriano's meditation on the Trinity, accompanied by the violas (Ex. 13.4), but they generate little dramatic intensity. The best things in the score are therefore the arias, where vocal grace and colourful instrumental writing are most effectively displayed.

Ex. 13.4

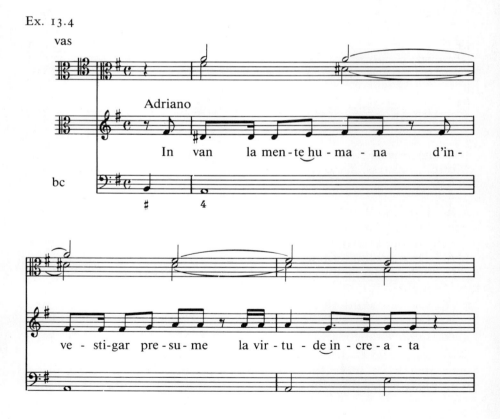

Ex. 13.4 *(cont.)*

di tre lu-mi di-stin-ti

(In vain the human mind presumes to investigate the uncreated strength of three separate lights.)

Perhaps only Scarlatti and Bononcini among Pistocchi's Italian contemporaries could rival the elegance of expression achieved in an aria like Adriano's 'Come lacrima il cigno dolente' (Ex. 13.5), in which the alto voice and solo cellos sing of death with such ardour, catching the whole spirit of the oratorio in one lyrical outburst.

Ex. 13.5

Ex. 13.5 (*cont.*)

Come la-cri-ma il ci-gno do-len-te   come

la-cri-ma il ci-gno do-len-te   quan-do   sen - te

(As the sad swan weeps when it feels . . .)

14

Vinacesi's *Susanna* (1694)

Susanna, an oratorio in two parts
Music by Cavaliere Benedetto Vinacesi
Libretto by Giovanni Battista Bottalino
Characters: Susanna   A.

Characters:	Susanna	A.
	Gioachino (Joachim)	S.
	Daniele (Daniel)	S.
	Primo Vecchio (First Elder)	T.
	Secondo Vecchio (Second Elder)	B.
	Chorus *a* 5	S. S. A. T. B.

Instrumental accompaniment *a* 5
First performed in Modena in 1694. In addition to the printed libretto in *I-MOe*, a manuscript libretto survives in *I-MOs* Archivio per Materie, Spettacoli Pubblici, Busta 1.

Benedetto Vinacesi (b. *c.*1670, Brescia; d. 1719, Venice) made his début in Modena in the same year as Pistocchi, 1692 (see Chapter 13), with an oratorio, *Gioseffo che interpreta i sogni*. Also in that same season, Gianettini revived Stradella's marvellous setting of Giardini's *Susanna* (see Chapter 4) which had greatly impressed the duke at its première in 1681. *Susanna*, like *S. Adriano*, is a story of matrimonial fedelity and may well have been performed on this second occasion to grace the duke's wedding celebrations. We have no way of knowing whether Vinacesi heard Stradella's oratorio in 1692, but two years later his own setting of the story was heard by the duke in what proved to be the last oratorio season of his reign. Little is known of the author, Bottalino; *Susanna* is his only extant oratorio.

The Libretto

Of Vinacesi's *Susanna*, Denis and Elsie Arnold have written: 'This really is an opera in all but name, and the example (if not the downright influence) of Stradella comes to mind.'[1] This appraisal rings true,

---

[1] See *The Oratorio in Venice* RMA Monographs, 2; (London, 1986), 18.

not only for the historical reasons given above, but more especially because Bottalino adopts a thoroughly operatic approach to the dramatization of the story. Like Giardini before him, he employs a cast of five characters and a chorus, but his choice of characters, and the importance he attaches to them, create an end-product quite unlike Giardini's oratorio. Bottalino employs a Chorus, but by restricting its role to the singing of a 'madrigal' at the very end of the oratorio, he removes the possibility of it having any moral impact on events. More drastic in effect is the replacement of Giardini's omniscient and authoritative Narrator with an extra dramatic role, that of Susanna's husband Joachim. This substitution dilutes the religious dimension of the oratorio and introduces a marital problem—a husband's mistrust of his chaste wife—into the proceedings to complicate the issues (a typical operatic ploy). With Susanna worrying as much about her husband as about the deceitfulness of the Elders, the only character championing the truth is the young prophet Daniel. He has to defend Susanna and expose the perjury of the Elders single-handed. In order to strengthen his hand, Bottalino has the good sense to allot an Introduction (an operatic prologue) to the Voice of Daniel, assuring the audience that the innocence of the chaste Susanna will be protected by the palm of victory crowning her head.

Having dispensed with a narrator, the author ensures that characters and changes of scene are signalled in the dialogue (cf. *S. Adriano*, Chapter 13).

The dramatic structure of the oratorio is as follows:

Sinfonia

Introduction	Voice of Daniel: The prophet asserts the moral strength of innocence. He describes the two lovelorn Elders hiding in Susanna's garden.
Part I:	
Scene 1	Elders, Susanna: The Elders lust after Susanna. They describe her undressing to bathe. Her happy song is picked up by an Echo, whose responses warn her: 'A snake . . . a snake . . . hidden . . . threatens . . . your honour.' She takes it as a sign from Heaven. Heaven will protect her.
Scene 2	Elders, Susanna: The Elders confront her. When she resists their overtures they threaten to denounce her as an adultress. She calls for help.
Scene 3	Joachim, Susanna, Elders: Joachim comes to her aid, but is confused by the malicious accusations of the

Elders. Outraged, he confines his wife to prison, pending a trial for adultery.

Scene 4   Susanna: She laments in prison her husband's loss of faith.

Part II:

Scene 5   Elders, Susanna, Joachim: The Elders call for the death penalty and the people prepare to stone Susanna. She and Joachim face death with fortitude.

Scene 6   Daniel, Elders: Daniel intervenes, speaking for the Spirit of God in declaring Susanna chaste. He examines the testimony of each Elder in turn, and proves them liars.

Scene 7   All: Joachim begs his wife's forgiveness. They are reconciled as the Elders are condemned to death.

Scene 8   All: In a concluding 'madrigal', Daniel records their death in a hail of stones. The Chorus points the moral, that a righteous God denies mercy to those who falsely accuse the innocent.

In Giardini's *Susanna*, the Judges, one an introvert, the other an extrovert, engaged in a diverting double-act for much of the oratorio. By comparison, Bottalino's Elders are lack-lustre. They are not differentiated in character and in the three arias they share (in Scenes 1, 2, and 5), mouth only platitudes. The Elders only come to life in the drama in passages of recitative dialogue.

From a historical standpoint, the most unusual feature of the libretto is the inclusion of an Echo in Susanna's monologue at the end of Scene 1 (cf. my remarks on the echoes in Palermino's *Sansone* in Chapter 11). From her aria 'Onde placide', the Echo picks up 'angue' (snake) from 'sangue' and 'langue'. Then from her recitative 'O sospirato', whole words from the ends of sentences reverberate: 'coperto' (hidden), 'insidia' (threatens), and 'l'onore' (your honour) to form an oracular pronouncement. With a nice touch of dramatic irony, Susanna thinks that the oracle is warning her of her husband's jealousy. She is unaware that two old snakes are lurking in the undergrowth, on the point of striking.

If one compares Bottalino's *Susanna* with its biblical source, The History of Susanna, it becomes clear that the operatic nature of the oratorio derives from the development of a dramatic role for Joachim. His relationship with Susanna, highly charged with conflicting emotions on both sides, forms a powerful sub-plot which, at times, almost obscures the main plot: the exposure and punishment of the

wicked Elders. As Daniel conducts the trial of the Elders, the proceedings are interrupted on three occasions by intimate confessions of
reconciliation between husband and wife, culminating in a sensuous
love-duet in Scene 7. The stark contrast between plot and sub-plot is
revealed at the end of the duet. Joachim, addressing his wife, sings,
'Peace to you, loving wife. Come to my arms!' Without further ado,
he addresses the audience in recitative: 'Thus, in his sin, perjured
injustice goes to his death because Heaven, which takes pleasure in
light, will not allow the sun to be obscured.' Here, the juxtaposition
of connubial bliss and divine severity is tasteless, a mixture of opera
and oratorio traditions that is an affront to both. It illustrates the real
dangers of trying to transform an Old Testament moral tale into a
'perfect spiritual melodrama'.

The Score

Vinacesi's setting of *Susanna* requires five solo voices; they combine to
sing the 'madrigal' finale *a 5*. The distribution of arias and ensembles
among the soloists is shown in Table 14.1. The casting of the female
title-role for alto voice, rather than soprano, is unusual for the period,
but a logical choice in the context of the full range of characters. The
imperious Daniel and the volatile Joachim have stronger claims to be
sung by castratos than the chaste Susanna.

TABLE 14.1. Distribution of Numbers in *Susanna*

	Daniel S.	Joachim S.	Susanna A.	2 Elders[a] T./B.
Continuo arias (2)	1	1	0	0
Arias with ritornellos (9)	2	2	3	2
Accompanied arias (8)	1	2	4	1
Duet with ritornellos (1)	0	1	1	0
Accompanied chorus (1)	1	1	1	1

[a] The Elders share three strophic arias.

The instrumental accompaniment *a 5* is for two violins, two alto
violas, and basso continuo. The Modenese copyist frequently writes
both viola parts on a single stave when they are providing simple
harmonic support to the more active violins above. There are some
passages of genuine five-part orchestral writing, but for much of the

score melodic interest is restricted to the two violins and continuo. For two of Susanna's arias the full orchestra is replaced by solo instruments: 'Caro amato' at the end of Part I is accompanied by a solo violin, and 'Vedervi à piangere' in Scene 7 by two cellos. In the latter aria, Vinacesi adopts precisely the same layout—alto voice, two cellos playing in the tenor clef, and basso continuo—as Pistocchi had used so effectively in Adriano's 'Come lacrima il cigno dolente' some two years earlier.

The opening sinfonia in E minor is a loosely constructed canzona in five contrasting sections, the first four of which end on an imperfect cadence:

A	Allegro: a fugal toccata in stile concitato	6 bars (repeated)
B	Grave mà spiccato: a homophonic section with chromatic harmony and roulades for the violin	11 bars
C	Presto: a fiery moto perpetuo for violin	16 bars
D	Presto: a fugue *a* 5	16 bars
E	Largo: a coda in triple time	8 bars

The last section completes the Sinfonia with a perfect cadence, but also functions as a ritornello for the first aria in the Introduction for the Voice of Daniel, 'Frange l'onde al mar turbato'. In this conventional simile-aria, the poet likens the strength of bare rocks resisting the turbulence of the waves to the strength of innocence fighting unarmed to defend itself against Fate. As the Sinfonia has carried us, without a break, into such thoughts, one may be justified in reading its stormy opening (Ex. 14.1) and fluctuating moods programmatically. A title like 'La Tempesta' springs to mind.

Bottalino's lyrics for arias invite da capo settings. The symmetry and sectionality of the poet's handiwork affects the musical settings. The first and second sections of Vinacesi's ternary structures are clearly defined, the da capos are exact repetitions (i.e. not written out) and, in many cases, symmetry is further enhanced by repeating an orchestral introduction as a coda. The da capo formula is also applied to the duet in Scene 7, leaving the final 'madrigal' as the only formal number matching the through-composed structure of the Sinfonia.

Though the form of Vinacesi's arias is predictable, each one is strongly characterized. The thunderous power of the orchestra, heard in the sinfonia, is called upon again as the Elders sing their aria denouncing Susanna in Scene 5, 'Caderà, quell'ingrata'. When the tables are turned, however, the stile concitato puts in a final appearance on the side of righteousness, supporting Daniel's aria 'Scenda nel popolo di Dio' (Ex. 14.2). As we have observed, solo obbligatos

Ex. 14.1

Ex. 14.2

(May [terrible wrath] fall on the people)

occur in two of Susanna's arias. In Scene 7, two cellos and continuo provide a gentle trio accompaniment to 'Vedervi à piangere' (Ex. 14.3); though the composer stipulates an affectionate performance ('affettuoso assai'), there is little in the simple imitative texture to stir the emotions. Much more effective is 'Caro amato', with violin solo (Ex. 14.4). Here, the imprisoned Susanna is recalling the delights of her marriage, and wondering what trick of fate has separated her from her husband. Vinacesi's expression mark is in the superlative form, 'affettuosissimo', and the notes of the violin's introduction match

Ex. 14.3

(To see you in tears, loveliest eyes)

Ex. 14.4

Ex. 14.4 *(cont.)*

Ex. 14.4 *(cont.)*

(Dear, beloved, my delight, my husband, what trickery steals you from me?)

that intention. The alto voice, not to be outdone, has an intricate chromatic melisma on 'inganno' (trickery), but can do nothing to compete with the astonishing passage of triple-stopping with which the violinist concludes the first section of the aria. The middle section is enlivened by a friendly exchange of trills between voice and violin to illustrate 'torna'. Each of the five voices has technically challenging music to perform, but the brunt of virtuoso passages is borne by the castrato, Joachim. His aria in Scene 3, 'Empie furie che m'agitate', contains an eleven-bar setting of 'flagellate' (Ex. 14.5), which is truly punishing.

Ex. 14.5

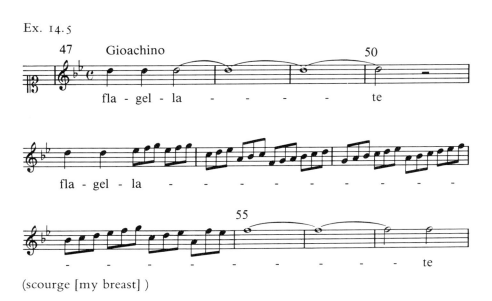

(scourge [my breast])

The ensembles occur at the very end of the oratorio. They are so deftly handled that they leave one wishing that the librettist had devised a few more in the earlier stages of the work. For the reconciliation duet of husband and wife, 'Pace, amata moglie | Pace, amato sposo', Vinacesi gives us smooth counterpoint in a lilting triple time. The rapture of their restored affection for each other is expressed in the flowing thirds and sixths of their 'Torna in questo sen' (Ex. 14.6).

The stern message of the final chorus calls for music of a grander quality. Massed voices and strings, in ponderous chordal style, demand the audience's attention: 'You, who are witnesses of this tragic scene' (i.e. the death of the Elders). The voice of Daniel leads into a solemn five-part vocal fugue: 'Understand this, that the just

Ex. 14.6

Ex. 14.6 *(cont.)*

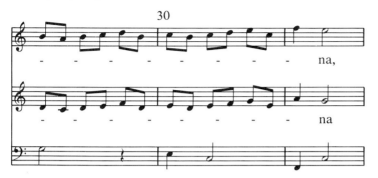

(Peace, peace, beloved wife. Come, turn . . .) ‖ Peace, peace, beloved spouse. Come, turn . . .)

God buries the wicked under a mountain of stones' (Ex. 14.7*a*). Again the orchestra joins the voices in harmony for 'In order to see rejected'. A more elaborate fugue on two subjects then ensues for 'he who for love offends and accuses the innocent'. Appropriately, the voice of Joachim leads the fugue on 'Chi pecca amando', and the voice of his wife that on 'e l'innocenza accusa' (Ex. 14.7*b*). After the voices alone

Ex. 14.7

a

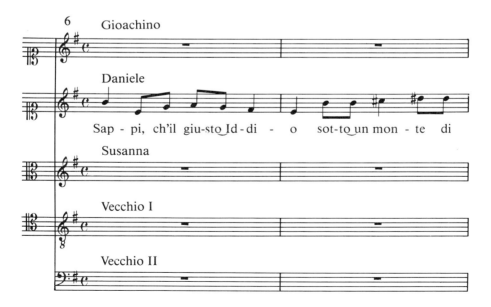

Ex. 14.7 *(cont.)*

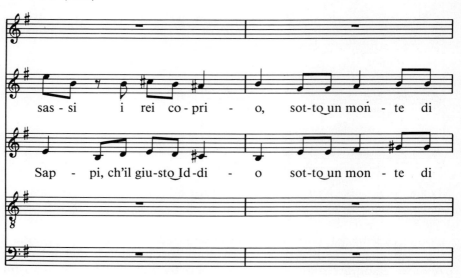

Ex. 14.7 (*cont.*)

Ex. 14.7 *(cont.)*

(Understand this, that the just God buries the wicked under a mountain of stones . . . He who for love offends and accuses the innocent.)

have sustained the double counterpoint for thirteen bars, the orchestra takes up the threads for a further seven bars. The music returns to the tonic key, E minor, as the first Elder leads the voices back into the fugue for an exhilarating concerted ending in nine real parts.

On the evidence of this splendid chorus, it is not surprising to find that Vinacesi's best works are considered to be the church music he wrote for San Marco in Venice in the early eighteenth century. There is much to admire in his oratorio, *Susanna*, but the genre in its modern form was too limited in scope for his particular talents.

15

Conclusions

There is still much to be discovered about the oratorio in Modena. As far as its early history is concerned, the discovery of the whereabouts of Soliani's *Raccolta d'oratorii*, i, a collection of librettos from the 1670s, would greatly expand our knowledge of that decade, as would further research into the activities of the Theatines and the Congregazione di San Carlo. Primary source material for the 1680s and 1690s is more abundant, but even in this period many matters await investigation. One that springs immediately to mind is the network of cultural connections between Modena and Rome, through which a large number of Roman oratorios found their way to Modena. The answer to what these connections were may turn up in State papers, in family archives, or through close scrutiny of the Roman scores in the Biblioteca Estense.

Notwithstanding these areas of uncertainty, I offer a summary of my main findings, relating them, where possible, to developments in other oratorio centres.

In the 1680s Modena was undoubtedly one of Italy's most active centres of oratorio composition and performance. In the course of that decade, at least 57 oratorios were performed in the city (see Appendix 1). Such a rate of production, though falling well below that of the Oratorians in Rome and Florence where weekly performances throughout the winter season (from All Saints to Easter) were the norm, was higher than that of the oratory of San Marcello in Rome (50 oratorios) and way ahead of the oratories in Bologna (30) and Venice (22). Unfortunately, extant records give an incomplete picture of the previous decade in Modena, but the well-documented 1690s leave us in no doubt that the popularity of the genre declined towards the end of the century.

The main performance venue in Modena was the oratory of San Carlo rotondo, adjacent to the royal chapel of San Vincenzo and the monastery of the Theatines. This octagonal prayer-hall belonged to a lay fraternity, the Congregazione di San Carlo. It was furnished with a special tribune to accommodate the Este princes. Other performance venues, infrequently used, were the oratory of the Confraternità di San Geminiano, next to the church of San Giovanni del Cantone, the oratory of SS. Annunciata in the western quarter of the city, and the Teatro di Corte.

The bulk of oratorios was performed in Lent, though the survival of librettos dated November and May shows that the season was not confined to the six weeks before Easter. In Duke Rinaldo's reign oratorios were performed on State occasions, regardless of the time of year.

The business of oratorio production in Modena was a state enterprise. Historically, Francesco II was the first Italian head of state to follow the example of Emperor Leopold I in Vienna, and promote regular seasons of oratorio for his court. The high quality of the Cappella Ducale, the excellent credentials of successive maestri to oversee developments, and the involvement of the court's most talented poet, Giardini, in the enterprise, ensured its success.

The annals of Modenese oratorio (1665–1702) in Appendix 1 show that 114 performances were given of 83 individual works. The Este music collection contained a further 26 scores of oratorios, some of which may have been given an airing. Confirmatory evidence, however, in the form of extant librettos or archival documentation, is lacking. For the present, it is best to assume that these oratorios, like many of the opera scores in the Este collection, were acquired to enhance the duke's library, and perhaps provide a resource for his private diversion. If we categorize the 83 works for which there is evidence of public performance by the cities in which their composers were employed, then we find that 23 originated in Modena, 21 in Rome, 20 in Bologna, 6 in Ferrara, and 6 in Venice. The remainder came from Genoa, Milan, Mantua, Parma, Pistoia, and Brunswick. It is evident that the duke had good connections throughout northern Italy. Furthermore, his reputation as a generous patron of oratorio was recognized by the finest composers of the day. Though we cannot be sure of the year of composition of every oratorio performed in Modena, virtually all were composed after Francesco II came to power in 1674. His taste for oratorio, reflected in the repertory, was thus both cosmopolitan and up-to-date.

The quality of the oratorios produced in Modena ranged over the whole gamut. Having examined ten of the best (in Chapters 3, 4, 6–8, and 10–14), I should caution that, at the other end of the scale is work of astonishing crudity and naïvety: the score of Ferri's *La Vergine annonciata*, and Gargieria's libretto for Bazani's *La caduta di Gerico*. The sheer diversity of the repertory dispels any notion that there was a Modenese 'school' of oratorio composition. It is undeniable that Giardini and the court maestri made important contributions to the repertory, but the poet's influence on other Modenese librettists appears to have been slight, and the most prolific composer, Gianettini, wrote in a Venetian style, if not too outmoded, certainly too idiosyncratic to inspire the likes of Perti, Bononcini, or Pistocchi to follow him.

   The style and form of Italian music changed considerably in the last quarter of the seventeenth century, the informality of the Middle Baroque giving way to the formalism of the High Baroque. The canzona was superseded by the concerto and sonata, the dance anthology by the sonata da camera, the intrigue drama by the opera seria. Oratorio was as deeply affected by these changes as opera, and in Modena's cosmopolitan repertory we have a good opportunity to observe their impact. Among the composers at work in Modena, Ferrari and Vitali espoused Middle Baroque ideals. Colonna was a transitional figure (like Pasquini, Stradella, and de Grandis), adept at writing in the concerto style but reluctant to abandon the continuo aria and the variety of aria forms he had mastered in his youth. Scarlatti showed the way forward to the younger generation of composers: Pistocchi, Bononcini, and Vinacesi. Their oratorios show a bold expansion of lyric forms, firmly based on symmetrical, da capo, design.

   At the conservative and progressive ends of the spectrum were the septuagenarian Ferrari and the 20-year-old Bononcini. Ferrari's *Sansone* (1680) is a short oratorio, 768 bars long. Bononcini's *La Maddalena à piedi di Cristo* (1690) is huge by comparison, its 3,081 bars taking two and a half hours to perform. There are 18 arias in *Sansone*, of which 5 are accompanied by the orchestra; in *Maddalena* there are 43, of which about the same proportion (13) are accompagnato. It was not until the eighteenth century that the orchestra supplanted the continuo as the normal mode of accompaniment for arias in Italian oratorios. Less than half of Ferrari's arias are in ternary form whereas, with Bononcini, the da capo form is used in 75 per cent of the arias. Finally, if one compares the proportion of bars in each work allotted to recitative (the dramatic element) and aria (the lyric element), the balance in *Sansone* of 34 per cent recitative to 51 per cent aria is tilted in *Maddalena* in favour of the lyric form: 13 per cent recitative to 77 per cent aria. Between these two extremes lie the other 81 oratorios in the Modenese repertory, each with its own mixture of conservative and progressive elements. One further observation, based on a large sample of Modenese oratorios, is worth recording: the da capo aria, though skilfully developed by composers like Colonna and Gianettini in the 1680s, did not become the dominant form of aria until the 1690s.

   The range of subject-matter treated in Modenese oratories was typical of the late seventeenth century. Most popular were stories from the Old Testament (36 oratorios), closely followed by legends of the saints (31). Of the latter, 12 were tales of martyrdom, whilst five commemorated the heroic deeds of the Estensi. Less popular among Roman Catholics, perhaps because of their associations with Protestant dogmatics, were stories from the New Testament; only 13 settings from this source were presented in Modena. Though the morality play, a fourth source of material for oratorio, was success-fully promoted in Venice and Ferrara in the late seventeenth century, it was not a source of attraction for the members of the Congregazione di San Carlo. As far as we can tell, they only subjected themselves to fully fledged moralities on four occasions.

Many oratorios drawn from approved religious sources underwent subtle changes of emphasis in the process of dramatization. Giardini's *Susanna* (see Chapter 4) was not the only libretto highlighting the sensual attraction of a female protagonist: 13 others in the repertory testify to the popularity of the oratorio erotico in Modena. In cultivat-ing this particular subspecies of oratorio, the Modenese seem to have conducted themselves with greater self-restraint than some patrons of oratorio in Rome or Bologna. There are no records of scandals, closures, or arrests in Modena. Of course, that may have been a consequence of Duke Francesco's regular attendance at San Carlo rotondo, his presence a curb on indecorum. Another important con-sequence of the duke's close association with the oratory was that as many as 17 oratorios, though ostensibly religious in subject-matter, were transformed by their librettists into vehicles for political dis-course or even, at times, propaganda. Giardini's cycle of oratorios, *La vita di Mosè*, contains many examples of intrusive political arguments, related to affairs of State close to the duke's heart. The five dynastic oratorios, trumpeting paeans of praise to the blood royal, are at best, fervent expressions of loyalty, at worst, flagrant propaganda.

Viewed as a regional phenomenom within a complex religio-political history, the Modenese oratorio's growth in popularity can be seen to have coincided with the rising aspirations of the Este dynasty. It was at its most flourishing phase when the duke of Modena's sister became queen of England, and it began to decline almost immediately after the hopes invested in the Stuart alliance were dashed by the English Revolution of 1688. In cataloguing its decline, one can cite a few local factors which affected it, like the young duke's failing health, serious economic problems facing the Modenese exchequer, and a change of regime in 1695, but I would see the prime cause as the loss of morale brought on by what de Grandis described as 'the great disgrace that has befallen their Majesties of England'. This political catastrophe in a distant island sounded the death-knell of the oratorio in Modena.

APPENDIX I

A Chronological List of Oratorios performed in Modena, 1665–1702

Year of Performance	No.	Title[a]	Libretto		Score		1st Performance		Other Information
			Poet	Location	Composer	Location	Date	Venue[b]	
1665?	1	San Valeriano/Il battesimo di San Valeriano martire	Don M. Erculei	I-MOe	A. Paino	lost	Feast of St Cecilia, 22 Nov.	SC	
1677	2	San Contardo d'Este/Il senso depredato nell'abbandono del mondo dal gloriosissimo S. Contardo d'Este	Don G. G. Manzini	I-MOe	Don A. Ferrari	I-MOe Mus. F. 374			
1677	3	S. Antonio Abbate/..., l'eroe trionfator dell'Inferno	Don V. Carli	I-MOe	Don A. M. Pacchioni	lost	14 Feb.	SC	
1678	4	S. Ignazio/Le porpore trionfale del S. Martire Ignazio, il patriarca antiocheno	Don V. Carli	I-MOe	Don A. M. Pacchioni	I-MOe Mus. F. 842	1 Feb.		
1680	5	Oratorio di Gioseppe/Giuseppe Ebreo	Wasserman ('austriaco')	I-MOe	Don A. Ferrari	I-MOe Mus. F. 1557			
1680	6	Il Sansone	G. B. Giardini	I-MOe	Don B. Ferrari	I-MOe Mus. G. 74	Apr.	SC	
1680	7	Il Davidde/Il trionfo della Penitenza	Conte G. B. Rosselli	I-MOe	Don A. Ferrari	I-MOe Mus. F. 1550	1 May		
1681	8	Il transito di S. Gioseppe	G. A. Bergamori	I-MOe	G. P. Colonna	I-MOe Mus. F. 298		SC	
1681	9	L'Anima/Il trionfo dell'Anima[c]	anon.	I-MOe	G. M. Martini	I-MOe Mus. F. 701			
1681	10	La Susanna	G. B. Giardini	I-MOe	A. Stradella	I-MOe Mus. F. 1137	16 Apr.	SC	

Continued

Year of Performance	No.	Title[a]	Libretto		Score		1st Performance		Other Information
			Poet	Location	Composer	Location	Date	Venue[b]	
1682	11	*Il nascimento di Mosè (La vita di Mosè, 1)*	G. B. Giardini	I-MOe	Don V. de Grandis	I-MOe Mus. F. 519	May		
1682	12	*La gran' Matilde d'Este*	A. Colombo	I-MOe	Don A. M. Pacchioni	I-MOe Mus. E. 175			
1684	13	*L'Assalone*	G. A. Bergamori	I-MOe and A-Wn	G. P. Colonna	F-Pc D. 2302 and A-Wn 17697			
1684	14	*Giudith*	G. A. Bergamori	I-MOe	G. P. Colonna	lost		SC	
1684	15	*Il matrimonio di Mosè/La ritirata di Mosè dalla corte d'Egitto e suoi sponsali con Sefora (La vita di Mosè, 2)*	G. B. Giardini	I-MOe	Don V. de Grandis	I-MOe Mus. F. 521			Score dated 5 Feb. 1684
1684	16	*S. Edita/Oratorio di S. Edita, vergine e monaca, regina d'Inghilterra*	Don L. Orsini	I-MOe	A. Stradella	I-MOe Mus. F. 1142		SC	
1685	17	*La Maddalena pentita*	A. Colombo	I-MOe	A. Gianotti	I-MOe Mus. F. 506	Tues. of Holy Week	SG	
1685	18	*Il trionfo della morte*	anon.	I-MOe	B. Aleotti (Padre Palermino)	I-MOe Mus. F. 887			
1685	19	*S. Agnese*	B. Panfilio	I-MOe	B. Pasquini	I-MOe Mus. F. 907			
1685	20	*Abramo vincitor de' proprii affetti/Agar scacciata*	G. Malisardi	I-MOe	G. A. Perti	I-MOe Mus. F. 925			

Continued

Year of Performance	No.	Title[a]	Libretto		Score		1st Performance		Other Information
			Poet	Location	Composer	Location	Date	Venue[b]	
1685	21	Il Mosè conduttore del popolo ebreo (*La vita di Mosè*, 4)	G. B. Giardini	I-MOe	G. A. Perti	I-MOe Mus. F. 923		SC	
1685	22	S. Teodosia/ Il martirio di S. Teodosia	anon.	I-MOe	A. Scarlatti	I-MOe Mus. F. 1058 and 1059			
1686	23	L'innocenza depressa	anon.	I-MOe	P. degli Antonii	I-MOe Mus. F. 22		SC	
1686	24	La profezia d'Eliseo nell'assedio di Samaria	Dr G. B. Neri	I-MOe	G. P. Colonna	I-MOe Mus. F. 333		SC	
1686	25	Il Mosè legato di Dio/ . . . e liberator del popolo ebreo (*La vita di Mosè*, 3)	G. B. Giardini	I-MOe	G. P. Colonna	I-MOe Mus. F. 299	13 Mar.	SC	
1686	26	L'innocenza di Davide/ . . . illesa dai furori di Saullo	Marchese F. Sacrati	I-MOe	C. A. Lonati	I-MOe Mus. F. 640		SC	
1686	27	La Maddalena/Il trionfo della grazia overo La conversione di Maddalena	B. Panfilio	I-MOe	A. Scarlatti	I-MOe Mus. F. 1056		SC	
1686	28	L'ambitione debellata overo La caduta di Monmouth	G. A. Canal	I-MOe	G. B. Vitali	I-MOe Mus. E. 247		SC	
1686	29	La Giuditta	anon.	I-MOe	M. A. Ziani	F-Pc D. 13351		SC	

Continued

Year of Performance	No.	Title[a]	Libretto		Score		1st Performance		Other Information
			Poet	Location	Composer	Location	Date	Venue[b]	
1687	30	Primo e secondo miracolo di S. Antonio	anon.	I-MOe	P. S. Agostini	I-MOe Mus. E. 3		SC	
1687	31	La vittoria di Davide contra Golia	P. P. Seta	I-MOe	G. Bononcini	lost		SC	
1687	32	Amore alle catene / Miracolo terzo di S. Antonio	anon.	I-MOe	A. Gianettini	I-MOe Mus. F. 504		SC	
1687	33	L'Huomo in bivio	anon.	I-MOe	A. Gianettini	I-MOe Mus. F. 500		SC	
1687	34	S. Dimna / . . . figlia di rè d'Irlanda	G. A. Lorenzani	I-MOe	F. Lanciani	I-MOe Mus. F. 618		SC	
1687	35	Il sacrificio d'Abelle	B. Panfilio	I-MOe	A. Melani	I-MOe Mus. F. 731		SC	
1687	36	S. Alessio	P. F. Bernini	I-MOe	B. Pasquini	I-MOe Mus. F. 903		SC	
1687	37	Il martirio dei Santi / . . . Vito, Modesto e Crescenzia sotto la tirannide di Diocleziano imperatore	Abbate F. Contini	I-MOe	B. Pasquini	I-MOe Mus. F. 909			
1687	38	I fatti di Mosè nel deserto (La vita di Mosè, 5)	G. B. Giardini	I-MOe	B. Pasquini	I-MOe Mus. F. 908		SC	
1688	39	Il Giosue	T. Stanzini	I-MOe	G. Bononcini	I-MOe Mus. F. 103		SC	
1688	40	La caduta di Gierusalemme	G. A. Bergamori	I-MOe	G. P. Colonna	F-Pc D. 2303	Apr.	SC	

Continued

Year of Performance	No.	Title[a]	Libretto		Score		1st Performance		Other Information
			Poet	Location	Composer	Location	Date	Venue[b]	
1688	41	La creatione de' magistrati (La vita di Mosè, 6)	G. B. Giardini	I-MOe	A. Gianettini	I-MOe Mus. F. 501	4 Apr.	SC	Score dated 1688
1688	42	S. Maria Maddalena dei Pazzi	B. Panfilio	I-MOe	G. L. Lulier	I-MOe Mus. F. 671		SC	
1688	43	Loth	anon.		E. Millanta	I-MOe Mus. F. 746			
1688	44	S. Rosalia	anon.	I-MOe	B. Aleotti	I-MOe Mus. F. 885		SC	
1688	45	Il Sansone	anon.	I-MOe	B. Aleotti	I-MOe Mus. F. 886		SC	
1688	46	Il trionfo della castità	Dr G. M. Giannini	I-MOe	C. Pallavicino	I-MOe Mus. F. 895			
1688	47	S. Pelagia	anon.	I-MOe	A. Stradella	I-MOe Mus. F. 1127		SC	
1688	48	S. Giovanni Battista	Abbate G. Ansaldi	I-MOe	A. Stradella	I-MOe Mus. F. 1136		SC	
1689	49	La Bersabea	M. Bruquerez	I-MOe	Don G. d'Alessandri	I-MOe Mus. F. 15		SC	
1689	50	Il Giona	Don A. Ambrosini	I-MOe	G. B. Bassani	I-MOe Mus. F. 61		SC	
1689	51	La Vergine annonciata	Dr A. Vecchi	I-MOe	N. M. Ferri	I-MOe Mus. F. 380	25 Mar.	SC	
1689	52	Il martirio di S. Felicità	Marchese F. Sacrati	I-MOe	D. Gabrielli	lost		SC	
1689	53	S. Sigismondo, rè di Borgogna	D. Bernardoni	I-MOe	D. Gabrielli	I-MOe Mus. F. 426		SC	
1689	54	Abigaille	F. Bambini	I-MOe	B. Gaffi	lost		SC	
1689	55	La conversione della Beata Margherita da Cortona	G. B. Giardini	I-MOe	A. Gianettini	lost		SC	

Continued

Year of Performance	No.	Title[a]	Libretto		Score		1st Performance		Other Information
			Poet	Location	Composer	Location	Date	Venue[b]	
1689	56	Il martirio di S. Giustina	Marchese F. Sacrati	I-MOe	A. Gianettini	lost		SC	
1689	57	Il constituto di Cristo	Dr F. Torti	I-MOe	A. Gianotti	I-MOe Mus. F. 505		SC	
1689	58	S. Beatrice d'Este	B. Panfilio	I-MOe	G. L. Lulier	F-Pc D. 7217			
1689	59	La verginità trionfante/ . . . nelle purissime nozze di Maria sempre vergine col castissimo S. Giuseppe	Conte G. B. Rosselli	I-MOe	G. M. Martini	I-MOe Mus. F. 702	19 Mar.	SC	
1689	60	La sete di Cristo	anon.	I-MOe	B. Pasquini	I-MOe Mus. F. 905		SC	
1689	61	Il Giona	Abbate D. Bartoli	I-MOe	G. B. Vitali	I-MOe Mus. F. 1260		SC	
1690	62	La Maddalena à piedi di Cristo/ Maddalena in casa del Fariseo	L. Forni	I-MOe	G. Bononcini	I-MOe Mus. F. 102			
1690	63	L'Assalone							Repeat of 13
1690	64	La Micol	anon.	I-MOe	B. Gaffi	I-MOe Mus. F. 432 A-Wn 17684			
1690	65	La morte di Cristo/ La vittima d'Amore	Dr F. Torti	I-MOe	A. Gianettini				
1690	66	Abramo in Egitto	anon.	I-MOe	G. B. Viviani	I-MOe Mus. F. 1271			
1691	67	La forza del divino amore	P. Ottoboni	I-MOe	B. Gaffi	I-MOe Mus. F. 431			
1691	68	Dio sul Sinai (La vita di Mosè, 7)	G. B. Giardini	I-MOe	A. Gianettini	lost		SA	

Continued

Year of Performance	No.	Title[a]	Libretto		Score		1st Performance		Other Information
			Poet	Location	Composer	Location	Date	Venue[b]	
1691	69	Lo scisma del sacerdozio (La vita di Mosè, 8)	G. B. Giardini	I-MOe	A. Melani	lost			
1691	70	S. Rosa di Viterbo	anon.	I-MOe	A. Melani	F-Pc D. 7863			
1691	71	Oratorio di S. Giovanni Battista	anon.	I-MOe	E. Millanta	I-MOe Mus. F. 1552			
1692	72	Il fasto depresso / . . . nell'humiltà esaltata: oratorio per S. Edoardo II, rè d'Inghilterra	M. Pallai	I-MOe	C. Monari	I-MOe Mus. F. 760			
1692	73	I fatti di Mosè							Repeat of 38
1692	74	Il martirio di S. Adriano	S. Stampiglia	I-MOe	F. A. Pistocchi	I-MOe Mus. F. 947			
1692	75	S. Edita							Repeat of 16
1692	76	La Susanna							Repeat of 10
1692	77	Gioseffo che interpreta i sogni	Dr G. B. Neri	I-MOe	B. Vinacesi	lost			
1693	78	La Passione	C. Arnoaldi	I-MOe	A. Ariosti	I-MOe Mus. F. 23	6 Mar.		
1693	79	La caduta di Gerico	A. Gargieria	I-MOe	F. M. Bazani	I-MOe Mus. F. 63	Feb.		
1693	80	Davide penitente	Padre Conti	I-MOe	G. F. Garbi	I-MOe Mus. F. 473			
1693	81	La conversione della B. Margherita							Repeat of 55
1693	82	Gefte	Dr G. B. Neri	I-MOe	A. Gianettini	lost		TC	Performed in costume
1693	83	S. Pelagia							Repeat of 47

Continued

Year of Performance	No.	Title[a]	Libretto		Score		1st Performance		Other Information
			Poet	Location	Composer	Location	Date	Venue[b]	
1694	84	Il trionfo della fede ne' sponsali di Sofronia e di Olindo	anon.	I-MOe	S. Cherici	F-Pc D. 1952			
1694	85	Giuliano apostata	A. Gargieria	I-MOe	G. P. Colonna	lost	25 Mar.		
1694	86	La santità freno alla tirannide	anon.	I-MOe	A. Gangiura	I-MOe Mus. F. 472			
1694	87	Susanna	G. B. Bottalino	I-MOe	B. Vinacesi	I-MOe Mus. F. 1230			
1695	88	Il martirio di S. Giustina							Repeat of 56
1695	89	La morte di Cristo							Repeat of 65
1695	90	Dio sul Sinai							Repeat of 68
1696	91	La morte delusa[d]	anon.		G. B. Bassani	I-MOe Mus. F. 60			
1696	92	La vittoria di Davide							Repeat of 31
1696	93	La creatione de' magistrati							Repeat of 41
1696	94	Dio sul Sinai							Repeat of 68
1696	95	I fatti di Mosè							Repeat of 38
1696	96	S. Adriano							Repeat of 74
1696	97	Gefte					Dec.		Repeat of 82
1696	98	S. Agostino/Le finezze della divina grazia nella conversione di S. Agostino	Dr F. Torti	I-MOe	A. Gianettini	lost			
1697	99	S. Sigismondo							Repeat of 53
1697	100	S. Beatrice d'Este							Repeat of 58
1697	101	S. Agostino							Repeat of 98

Appendix 1

Continued

Year of Performance	No.	Title[a]	Libretto		Score		1st Performance		Other Information
			Poet	Location	Composer	Location	Date	Venue[b]	
1698	102	S. Agostino							Repeat of 98
1698	103	Gefte							Repeat of 82
1698	104	S. Giustina							Repeat of 56
1699	105	Gefte							Repeat of 82
1699	106	S. Agostino							Repeat of 98
1699	107	La conversione della B. Margherita							Repeat of 55
1699	108	S. Beatrice d'Este							Repeat of 58
1699	109	Il Giona							Repeat of 61
1699	110	S. Adriano							Repeat of 74
1700	111	La Maddalena à piedi di Cristo							Repeat of 62
1700	112	La morte di Cristo							Repeat of 65
1701	113	S. Beatrice d'Este							Repeat of 58
1702	114	La Micole							Repeat of 64

[a] More than one title is given if the score, the libretto, and archival listings differ markedly from each other

[b] Key to venues: SA = oratory of SS. Annunciata
SC = oratory of San Carlo rotondo
SG = Confraternità di San Geminiano
TC = Teatro di Corte

[c] Martini's *Le lacrime di S. Pietro* is bound with *L'Anima* in Mus. F. 701, but no libretto for it exists.

[d] The score is dated 1696; no Modenese libretto survives

Additions: Since Appendix 1 was compiled, evidence of two additional performances in Modena has come to light. Sebastiano Cherici's *La Santissima Annunciata*, listed in Appendix 2 as an additional oratorio owned by Francesco II, was performed in 1684. The title of a printed libretto bearing that date is among items listed in E. Milano, *Lavori preparatori per gli annali della Tipografia Soliani* (Modena, 1986), 23 ff. Giovanni Marco Martini's *Il disfacimento di sisara* (text by Count Filippo Maria Sartorio) was performed in 1693 according to a libretto in I-MOe. This oratorio is also listed in Appendix 2 as an additional oratorio owned by Francesco II.

APPENDIX 2

TABLE A. Additional Oratorios owned by Duke Francesco II (scores only)

Title	Poet	Composer	Location	Luin cat. no.[a]
Il ballo d'Erodiade	anon.	F. Collinelli	lost	80
La caduta d'Adamo	Cav. Nencini	Don V. de Grandis	I-MOe Mus. F. 518	8
Il disfacimento di Sisara	anon.	G. M. Martini	lost	90
Ester	anon.	A. Stradella	I-MOe Mus. F. 1155	15
Euridano, miracolo di S. Antonio	anon.	D. Freschi?	I-MOe Mus. F. 1546	97
Ezzelino	anon.	A. Boretti	I-MOe Mus. F. 116	96
Il giglio del Carmelo	anon.	G. A. Lorandi	F-Pc Rés. 1358	56
La Maddalena, piccolo oratorio	anon.	anon.	I-MOe Mus. E. 308	30
Il martirio di S. Eustachio	Card. P. Ottoboni	F. Lanciani	F-Pc Rés. 1359	74
Il miracolo del Mago	anon.	D. Freschi	I-MOe Mus. F. 393	98
Oratorio d'Abramo	anon.	anon.	I-MOe Mus. F. 1541	82
Il sacrificio d'Isaach	anon.	A. Martini	I-MOe Mus. G. 269	29
Il Salomone amante	G. A. Bergamori	G. P. Colonna	F-Pc D. 2304	3
S. Caterina martire	anon.	G. B. Gigli	I-MOe Mus. F. 508	70
S. Giovanni Grisostomo	anon.	A. Stradella	I-MOe Mus. F. 1132	14
S. Teodora	anon.	D. Leporati	lost	91
La Santissima Annunciata	anon.	S. Cherici	F-Pc D. 2227	4
Il Sisara	anon.	S. Martelli	I-MOe Mus. F. 696	23
La Susanna	anon.	G. B. Borri	F-Pc D. 1396	69
Il trionfo della pace per S. Filippo Benizzi	anon.	S. Cherici	F-Pc D. 1953	59
La tromba della divina misericordia	anon.	G. B. Bassani	I-MOe Mus. G. 14	22
La vendita del cor humano	P. M. Petrucci	G. Legrenzi	I-MOe Mus. F. 1544	39

[a] The numbers are from the listing of oratorios in E. J. Luin, 'Repertorio dei libri musicali di S. A. S. Francesco II d'Este', *Bibliofilia*, 38 (1936), 418–45.

TABLE B. Additional Oratorios owned by Duke Rinaldo I (scores only)

Title	Poet	Composer	Location
Gilard ed Eliada	anon.	Padre A. F. Urio	*I-MOe* Mus. F. 1200
S. Ermenegildo	anon.	P. Magagni	*I-MOe* Mus. F. 674
S. Genovefa palatina	anon.	G. B. Gigli	*I-MOe* Mus. F. 509
Il viaggio di Tobia	anon.	G. M. Casini	*I-MOe* Mus. F. 147

SELECT BIBLIOGRAPHY

For a comprehensive bibliography of Italian oratorio in the Baroque era, see
H. E. Smither, *A History of the Oratorio*, i (Chapel Hill, NC, 1977),
435–63. The following books and articles relate to the oratorio in Modena.

ALLAM, E., 'Alessandro Stradella', *Proceedings of the RMA*, 80 (1954), 29–42.

AMORTH, L., *Modena capitale: storia di Modena e dei suoi duchi dal 1598 al 1860*
(Milan, 1967).

ARNOLD, D. and E., *The Oratorio in Venice* (RMA Monographs, 2; London,
1986).

BECHERINI, B., 'Dal Barocco all'oratorio di G. Bononcini', *Chigiana*, 13
(1956), 97–110.

BELTRAMI, G., 'Il ducato di Modena tra Francia e Austria (Francesco II
d'Este, 1674–1694)', *Atti e memorie, Deputazione di Storia Patria per le antiche
provincie modenese*[8], 9 (1957), 100–42.

CALORE, M., *Spettacoli a Modena tra '500 e '600* (Modena, 1983).

CASIMIRI, R., 'Oratorii del Masini, Bernabei, Melani, Di Pio, Pasquini e
Stradella', *Note d'archivio per la storia musicale*, 13 (1936), 157–69.

CATELANI, A., 'Delle opere di Alessandro Stradella esistenti nell'archivio
musicale della R. Biblioteca Palatina di Modena', *Atti e memorie delle R. R.
Deputazioni di Storia Patria per le provincie modenese e parmensi*, 3 (whole
issue) (1865).

CAVELLI, C. DE, *Les Derniers Stuarts à Saint-Germain-en-Laye* (Paris, 1871).

CHIAPPINI, L., *Gli Estensi* (Milan, 1967).

CHIARELLI, A., *I codici di musica della Raccolta Estense* (Florence, 1987).

CROWTHER, J. V., 'The Development of Oratorio in Emilia, 1650–1700',
Ph.D. thesis (Nottingham, 1977).

—— 'Alessandro Stradella and the Oratorio Tradition in Modena', in
Gianturco (ed.), *Alessandro Stradella e Modena*, 51–64.

—— 'The Characterization of Women in Stradella's Oratorios', *Chigiana*, 39
(1988), 277–85.

—— 'A Case-Study in the Power of the Purse: The Management of the
Ducal *Cappella* in Modena in the Reign of Francesco II d'Este', *Journal of
the RMA*, 115: 2 (1990), 207–19.

DAVID, C., *Vive Jesus* (Aix-en-Provence, 1670). Rare copy, *GB-LbL* 1484
ee4.

ERCULEI, M., *Il canto ecclesiastico* (Modena, 1686).

GANDINI, A., *Cronistoria dei teatri di Modena dal 1539 al 1871* (Modena,
1874).

GIANTURCO, C., 'The Oratorios of Alessandro Stradella', *Proceedings of the
RMA*, 101 (1974–5), 45–57.

—— (ed.), *Alessandro Stradella e Modena* (Modena, 1985).

—— and ROSTIROLLA, G. (edd.) 'Alessandro Stradella e il suo tempo',

Chigiana, 39 (special issue in 2 vols.) (1988).

—— and McCRICKARD, E. (edd.), *Alessandro Stradella, 1639–1682: A Thematic Catalogue of his Compositions* (New York, 1991).

HESS, H., 'Die Opern Alessandro Stradellas', *Beihefte der Internationalen Musikgesellschaft*, 3 (1906).

JANDER, O., *Alessandro Stradella* (The Wellesley Edition cantata index series, 4; Wellesley, Mass., 1969).

—— 'The Cantata in Accademia: Music for the Accademia de' Dissonanti and their Duke, Francesco II d'Este', *Rivista italiana di musicologia*, 10 (1975), 519–44.

JOHNSON, J. L., and SMITHER, H. E. (edd.) *The Italian Oratorio, 1650–1800*, (Garland Press facs. repr.; New York, 1986), v, vi.

KLENZ, W., *Giovanni Maria Bononcini* (Durham, NC, 1962).

LODI, P., 'Catalogo delle opere musicali—città di Modena', *Bolletino dell'Associazione dei musicologia italiana* (1916–24).

LUIN, E. J., *Antonio Gianettini e la musica a Modena* (Modena, 1931).

—— 'Repertorio dei libri musicali di S. A. S. Francesco II d'Este nell'Archivio di Stato di Modena', *Bibliofilia*, 38 (1936), 418–45.

MAYLENDER, M., *Storie delle accademie d'Italia* (Bologna, 1926–30).

MISCHIATI, O., 'Per la storia dell'oratorio a Bologna: tre inventari del 1620, 1622, e 1682', *Collectanea historiae musicae*, 3 (1963), 131–70.

MURATORI, L. A., *Delle Antichità Estensi*, ii (Modena, 1740).

O'LOUGHLIN, N., 'Stradella's Santa Pelagia', *Musical Times* (May 1981), 297–300.

OMAN, C., *Mary of Modena* (London, 1962).

PASTOR, L., *The History of the Popes*, trans. Antrobus, Kerr, *et al.* (St Louis, 1891–1953).

Raccolta d'oratorii per musica fatti cantare in diversi tempi dall'Altezza serenissima di Franceso. II, Duca di Modona, Regio, ecc. nell'oratorio di S. Carlo di Modona (Soliani Press, Modena, 1689) ii. iii. Only copies in *I-MOe* 83-I-6 (vol. ii) and 83-I-5 (vol. iii). (vol. i lost.)

RONCAGLIA, G., 'Di insigni musicisti modenesi', *Atti e memorie della R. Deputazione di Storia Patria per le provincie modenese*[7], 6 (1929).

—— 'Giuseppe Colombi e la vita musicale modenese', *Atti e memorie dell'Accademia di Scienze, Lettere e Arti*[5], 10 (1952), 47–52.

—— 'La scuola musicale modenese', *Chigiana*, 13 (1956), 69–83.

—— *La cappella musicale del duomo di Modena* (Florence, 1957).

SCHENK, E., 'Osservazioni sulla scuola istrumentale modenese', *Atti e memorie dell'Accademia di Scienze, Lettere e Arti*[5], 10 (1952), 3–28.

SCHERING, A., *Geschichte des Oratoriums* (Leipzig, 1911).

SOLI, G., *Chiese di Modena* (Modena, 1974).

SOSSAJ, F., *Descrizione della città di Modena nell'1833* (Modena, 1833; repr. Bologna, 1972).

SPINELLI, A. G., 'Gio. Marco Martini, contrappuntista del secolo 17 alla corte estense', *Atti e memorie, Deputazione di Storia Patria per le antiche provincie modenese*[4], 4 (Modena, 1893), 211–14.

TARDINI, V., *I teatri di Modena* (Modena, 1902).

TIRABOSCHI, G., *Biblioteca modenese*, (Modena, 1781–6), i–vi.

—— *Notizie de' pittori scultori, incisori e architetti natii degli stati del Serenissimo Signor Duca di Modena, con un appendice de' professore di musica, raccolte dal cavaliere ab. G. Tiraboschi* (Modena, 1786).

VALDRIGHI, L. F., 'Cappelle, concerti e musiche estensi', *Atti e memorie della Deputazione di Storia Patria per le provincie modenese*[3], 2 (1883), 415–95.

VATIELLI, F., 'L'oratorio a Bologna', *Note d'archivio per la storia musicale*, 15 (1938), 26–35, 77–87.

WORSTHORNE, S. T., *Venetian Opera in the Seventeenth Century* (Oxford, 1953).

INDEX OF ORATORIOS

Items are presented in the following order: short title (poet) composer [no. in Appendix 1] main textual references.

Abigaille (Bambini) Gaffi [54] 117
Abramo in Egitto (anon.) Viviani [66]
Abramo vincitor (Malisardi) Perti [20] 62, 63–4
Adamo, L' (anon.) Cossoni 60
Agare (Mauritio) Vitali 20, 145
Ambitione debellata, L' (Canal) Vitali [28] 58, 66, 145
Amore alle catene (anon.) Gianettini [32] 111
Anima, L' (anon.) Martini [9] 25
Assalone, L' (Bergamori) Colonna [13] 61

Ballo d'Erodiade, Il (anon.) Collinelli 202
Beata Margherita, La (Giardini) Gianettini [55] 117
Bersabea, La (Bruquerez) Alessandri [49] 117

Caduta d'Adamo, La (Nencini) de Grandis 202
Caduta di Gerico, La (Gargieria) Bazani [79] 187
Caduta di Gierusalemme, La (Bergamori) Colonna [40] 114
Caino condannato, Il (Savaro) Cazzati 60
Constituto di Cristo, Il (Torti) Gianotti [57] 117
Creatione de' magistrati, La (Giardini) Gianettini [41] 60, 113, 121–32

Davidde, Il (Rosselli) Ferrari [7] 23–4
Davide penitente (Conti) Garbi [80]
Dio sul Sinai (Giardini) Gianettini [68] 60, 119
Disfacimento di Sisara, Il (anon.) Martini 200 n., 202

Ester (anon.) Stradella 202
Euridano (anon.) Freschi? 202
Ezzelino (anon.) Boretti 202

Fasto depresso, Il (Pallai) Monari [72] 119
Fatti di Mosè, I (Giardini) Pasquini [38] 60, 62, 110
Forza del divino amore, La (Ottoboni) Gaffi [67]

Gefte (Balbi) Vitali 20, 145
Gefte (Neri) Gianettini [82]
Giglio del Carmelo, Il (anon.) Lorandi 202
Gilard ed Eliada (anon.) Urio 203
Giona, Il (Ambrosini) Bassani [50] 117

Giona, Il (Bartoli) Vitali [61] 117, 125, 145–55
Gioseffo che interpreta i sogni (Neri) Vinacesi [77] 169
Giosue, Il (Stanzini) Bononcini [39] 112, 114
Giudith (Bergamori) Colonna [14] 61
Guiditta (anon.) Cazzati 20
Giuditta, La (anon.) Ziani [29] 67–8
Giuliano apostata (Gargieria) Colonna [85]
Giuseppe Ebreo (Wasserman) Ferrari [5] 23–4

Huomo in bivio, L' (anon.) Gianettini [33] 111

Innocenza di Davide, L' (Sacrati) Lonati [26] 67
Innocenza depressa, L' (anon.) Antonii [23] 66–7

Lacrime di S. Pietro, Le (anon.) Martini 200 n.
Loth (anon.) Millanta [43] 114

Maddalena, La (Panfilio) Scarlatti [27] 67, 81
Maddalena pentita, La (Colombo) Gianotti [17] 62, 65
Maddalena à piedi di Cristo, La (Forni) Bononcini [62] 119, 188
Maddalena, piccolo oratorio, La (anon.) anon. 202
Martirio di S. Adriano, Il (Stampiglia) Pistocchi [74] 119, 156–68, 169–70
Martirio di S. Eustachio, Il (Ottoboni) Lanciani 202
Martirio di S. Felicità, Il (Sacrati) Gabrielli [52] 117
Martirio di S. Giustina, Il (Sacrati) Gianettini [56] 117
Martirio di Santi, Il (Contini) Pasquini [37] 111
Matilde d'Este, La gran' (Colombo) Pacchioni [12] 59, 60
Matrimonio di Mosè, Il (Giardini) de Grandis [15] 60–1, 69
Micole, La (anon.) Gaffi [64] 117
Miracoli di S. Antonio (anon.) Agostini [30] 111
Miracolo del Mago, Il (anon.) Freschi 202
Morte delusa, La (anon.) Bassani [91] 120
Morte di Cristo, La (Torti) Gianettini [65]

Mosè conduttore, Il (Giardini) Perti [21] 60, 62, 63–4
Mosè legato di Dio, Il (Giardini) Colonna [25] 60, 66, 69

Nascimento di Mosè, Il (Giardini) de Grandis [11] 59, 60, 69–80, 125

Oratorio d'Abramo (anon.) anon. 202
Oratorio del diluvio (Savaro) Cazzati 60
Oratorio di S. Giovanni Battista (anon.) Millanta [71]

Passione, La (Arnoaldi) Ariosti [78]
Profezia d'Eliseo, La (Neri) Colonna [24] 66, 95–108

Sacrificio d'Abelle, Il (Panfilio) Melani [35] 111
Sacrificio d'Isaach, Il (anon.) Martini 202
Salomone amante, Il (Bergamori) Colonna 202
S. Agnese (Panfilio) Pasquini [19] 62–3, 81, 114
S. Agostino (Torti) Gianettini [98] 119–20
S. Alessio (Bernini) Pasquini [36] 111
S. Antonio Abbate (Carli) Pacchioni [3] 21–2
S. Azzo Estense (Giardini) anon. 22 n.
S. Beatrice d'Este (Panfilio) Lulier [58] 114, 117
S. Caterina martire (anon.) Gigli 202
S. Contardo d'Este (Manzini) Ferrari [2] 21–2
S. Dimna (Lorenzani) Lanciani [34] 111
S. Edita (Orsini) Stradella [16] 61
S. Ermenegildo (anon.) Magagni 203
S. Genovefa palatina (anon.) Gigli 203
S. Giovanni Battista (Ansaldi) Stradella [48] 45, 46, 113, 129
S. Giovanni Grisostomo (anon.) Stradella 202
S. Ignazio (Carli) Pacchioni [4] 22–3
S. Maria Maddalena dei Pazzi (Panfilio) Lulier [42] 114
S. Pelagia (anon.) Stradella [47] 113
S. Rosa di Viterbo (anon.) Melani [70]
S. Rosalia (anon.) Aleotti [44] 115–16, 133
S. Sigismondo (Bernardoni) Gabrielli [53] 117
S. Teodora (anon.) Leporati 202

S. Teodosia (anon.) Scarlatti [22] 62, 63, 81–94, 96, 157
S. Valeriano (Erculei) Paino [1] 18, 19
Sansone, Il (anon.) Aleotti [45] 115, 133–44, 171
Sansone, Il (Giardini) Ferrari [6] 24, 27–39, 125, 133, 155, 188
Santissima Annunciata, La (anon.) Cherici 200 n., 202
Santità freno alla tirannide, La (anon.) Gangiura [86]
Scisma del sacerdozia, Lo (Giardini) Melani [69] 60, 119
Sete di Cristo, La (anon.) Pasquini [60] 117
Sisara, Il (anon.) Martelli 202
Sospetto con il pianto (Ferrari) Ferrari 26
Sponsali d'Ester, Gli (anon.) Legrenzi 19–21
Susanna (Bottalino) Vinacesi [87] 119, 169–85
Susanna, La (anon.) Borri 202
Susanna, La (Giardini) Stradella [10] 26, 40–57, 70, 122–3, 169, 171, 189

Transito di S. Gioseppe, Il (Bergamori) Colonna [8] 25–6, 125
Transito di S. Gioseppe, Il (Sanuti) Cazzati 25
Trionfo della castità, Il (Giannini) Pallavicino [46] 114–15, 146
Trionfo della fede, Il (anon.) Cherici [84]
Trionfo della fede, Il (Tesini) Pratichista 145
Trionfo della morte, Il (anon.) Aleotti [18] 62, 64–5, 133, 146
Trionfo della pace, Il (anon.) Cherici 202
Tromba della divina misericordia, La (anon.) Bassani 202

Vendita del cor humano, La (Petrucci) Legrenzi 202
Vergine annonciata, La (Vecchi) Ferri [51] 116, 187
Verginità trionfante, La (Rosselli) Martini [59] 117
Viaggio di Tobia, Il (anon.) Casini 203
Vittoria di Davide, La (Seta) Bononcini [31] 112

GENERAL INDEX

This index contains the names of persons and institutions (religious and civic) mentioned in the text. Dramatis personae and biblical characters are excluded. The names of saints, members of the Este family, and churches are grouped under their respective headings. As spelling was inconsistent in the seventeenth century I have adopted those forms most commonly found in Modenese sources, e.g. 'Panfilio' rather than 'Pamphilii', showing alternative spellings in square brackets. Unless otherwise indicated, the institutions listed in the index are, or were, located in Modena. Page numbers in italics following the name of a composer or librettist signify a chapter devoted to an analysis (with music examples) of one of his oratorios.

Accademia dell'Arcadia (Rome) 156
Accademia de' Dissonanti 2, 25, 58, 65, 145
Accademia Filarmonica (Bologna) 112
Agatea, Maria 16
Agostini, Pier Simone 111, 195
Albarelli, Luigi 156
Albergati, Pirro 95
Albertini, Giacinto 16
Aldobrandini, cardinal Pietro 1
Aleotti, Bonaventura, *see* Palermino, padre
Alessandri, Giulio d' 117, 196
Alexander VIII, pope 63 n.
Allam, Edward 205
Allemani, Antonio 59, 118
Ambrosini, Ambrogio 196
Amorth, Luigi 205
Ansaldi, Giovanni 196
Antonii, Pietro degli 66–7, 194
archive of San Petronio (Bologna) 63 n.
archivio di stato, *see* State archive
Ariosti, Attilio 198
Ariosto, Ludovico 3
Arnoaldi, Camillo 198
Arnold, Denis 20 n., 115 n., 169, 205
Arnold, Elsie 20 n., 115 n., 169, 205
Arresti, Giulio Cesare 60
Ascani, Pellegrino 16, 118
Ascani, Simone 16
Avanzini, Bartolomeo 2

Baldrini, Giovanni Battista 110 n., 118
Balducci, Francesco 30
Ballerini, Francesco 64
Balugani, Antonio 16, 64
Bambini, Francesco 196
Baraoni, Giovanni Maria 16
Barberini family 12
Barbieri, Giulio 17, 118
Bartoli, Domenico 145–55, 197
basilica of San Marco (Venice) 68, 109, 185
basilica of San Petronio (Bologna) 18, 25, 95, 97, 98, 112

Bassani, Giovanni Battista 117, 120, 196, 199, 202
Bazani, Francesco Maria 187, 198
Becherini, Bianca 205
Belletti, Paolo 17
Bellini, Ippolito 17, 118
Beltrami, G. 207
Bendinelli, Agostino 23
Bergamori, Giacomo Antonio 25, 98, 114, 192, 193, 195, 202
Bernardoni, Domenico 196
Bernini, Gianlorenzo 3
Bernini, Pietro Filippo 195
Bianconi, Lorenzo 25 n.
Biblioteca Estense 6, 18, 61, 84, 186
Bononcini, Giovanni 5 and n., 6, 112, 114, 119, 156, 167, 188, 195, 197
Bononcini, Giovanni Maria 3, 5, 17, 23
Boretti, Antonio 202
Borri, Giovanni Battista 202
Borromeo, Carlo, *see* saints: Charles Borromeo
Borromini, Francesco 62
Boschetti, Paolo 2
Bottalino, Giovanni Battista *169–85*, 199
Bozzoleni, Giovanni 64
Braida, Giovanni 117
Bratti, Domenico 5, 17, 118
Bruquerez, Michele 196
Busca, Angelo Michele 15
Bussi, Steffano 16

Callegari, Laura 40
Calore, Marina 1 n., 205
Canal, Giovanni Andrea 194
Canori Press 6
Capiluppi, Francesco 17, 118
Cappella Ducale 4–6, 14–19, 40, 58, 110, 117–20, 187
Capuchin friars 10
Carissimi, Giacomo 95, 98
Carli, Valentino 21, 192

Carracci, school of 19 n.
Casanova, Giovanni Battista 17, 118
Casimiri, Raffaele 59 n., 205
Casini, Giovanni Maria 203
Cassiani Press 6, 18
Castlemaine, Lord 110
Catelani, Angelo 205
cathedral Chapter 6, 7
cathedral of San Geminiano xi, 3, 5, 6, 7, 11,
    17 n., 21, 59
cathedral of Santiago de Compostela (Spain)
    21
Cavalli, Francesco 15, 24, 109
Cavelli, Campana de 12 n., 65 n., 113 n., 205
Cazzati, Maurizio 20, 25, 60, 95
Cecchi, Domenico 64
Cerlini, Vittorio 16, 59
Cestellino 11
Cesti, Antonio 24
Charles II, king of England 12, 65
Cherici [Clerici], Sebastiano 199, 200 n., 202
Chiappini, Luciano 10 n., 205
Chiaravalli, Ferdinando 64
Chiarelli, Alessandra 26 n., 84 n., 205
Christina, queen of Sweden 62, 63, 81, 110,
    156
churches:
    Crocefisso xi, 2
    Gesù (Rome) 58
    Holy Trinity, *see* Santa Maria delle Assi
    Madonna di Galliera (Bologna) 25, 95
    Madonna del Paradiso, *see* Paradiso
    Paradiso xi, 2, 7, 8, 58
    Sant' Agnese (Rome) 62
    Sant' Agostino xi, 2, 10–11, 113
    Sant' Andrea della Valle (Rome) 8
    San Bartolomeo xi, 2
    San Biagio xi, 2
    San Carlo Borromeo xi, 2
    Santa Casa (Loreto) 6, 58, 113, 116
    San Geminiano, *see* cathedral of San
        Geminiano
    San Giorgio xi, 2
    San Giovanni del Cantone xi, 8, 186
    San Giovanni in Monte (Bologna) 6, 112
    San Leonardo (Fano) 109
    San Marco (Venice), *see* basilica of San
        Marco (Venice)
    Santa Maria delle Assi 23
    Santa Maria della Fava (Venice) 20 n.
    San Petronio (Bologna), *see* basilica of San
        Petronio (Bologna)
    San Pietro (Reggio) 26
    San Vincenzo xi, 2, 5, 8, 11, 109, 186
    Voto xi, 2
Ciocchi, Giovanni 17, 118
Clement VIII, pope 7

Clement X, pope 20
Coleman, Edward 12
College of Nobles xi, 2, 8
Collinelli, Filippo 202
Colombi, Giuseppe 5, 14, 16, 17 n., 18, 61,
    115, 117–19
Colombo, Alfonso 193
Colonna, Giovanni Paolo 5 n., 25, 26, 60, 61,
    66, *95–108*, 112, 114, 119, 125, 188,
    192–5, 199, 202
Compagnia dei Violini 4
Comune 1, 2, 6
Congregation of Bishops (Rome) 10
Congregation of the Oratory, *see* Oratorians
Congregazione della Beata Vergine e di San
    Carlo 2, 8–9, 13, 18, 19, 21, 25, 60, 62,
    111, 116, 186, 189
Conservatorio degl' Incurabili (Venice) 115
Conti, padre 198
Contini, Francesco 195
convent (Castelnuovo di Garfagnana) 10
convent of the discalced Carmelites xi, 2, 7,
    10
convent, Salesian xi, 2, 7, 10, 11
convent, Salesian (Aix-en-Provence) 10
convent, Theatine xi, 8, 9
Corelli, Arcangelo 110
Corregio of Mantua 156
Cossoni, Carlo Donato 60
Cottino, *see* Pietrogalli, Antonio
Crowther, John Victor 4 n., 60 n., 66 n., 205

Dante Alighieri 22
David, Charles 10 n., 205
Degni Press 18
discalced Carmelites, *see* convent of the
    discalced Carmelites
Donati, Giuseppe detto Il Tintorino 45, 59
Ducal Palace xi, 2, 5, 10
Ducal Palace (Sassuolo) 6, 109, 117

Emanuele Filiberto, prince of Savoy 65
English parliament 12, 116
Erculei, Marzio 14, 16, 18 and n., 19, 192,
    205
Este:
    Alessandro d', prince of Modena 3
    Alfonso II d', duke of Ferrara 1, 3
    Alfonso III d', duke of Modena 3, 8, 9, 10
    Alfonso IV d', duke of Modena 3, 4, 5, 10,
        14, 27
    Angela Maria Caterina d', princess of
        Modena 65
    Carlotta Felicita d', duchess of Modena 12
    Cesare d', duke of Modena 1, 3, 7
    Cesare Ignazio d', prince of Modena 11, 12,
        65, 110

Eleanora d', princess of Modena 10
Francesco I d', duke of Modena 2, 3, 4, 5, 10, 14
Francesco II d', duke of Modena 2–6, 9, 10–12, 14–20, 23 n., 24, 27, 40, 44, 57–62, 64–5, 69, 79, 109–15, 117–19, 156, 169, 187, 189, 200 n., 202
Isabella d', duchess of Modena 8
Laura Martinozzi d', duchess and regent of Modena 3–5, 10–11, 14, 15, 17 n., 19, 27, 109–13
Lucrezia Barberini d', duchess of Modena 119
Margherita Farnese d', duchess of Modena 119, 156
Maria Beatrice d', princess of Modena and queen of England 10–12, 22, 58, 61, 65–6, 110, 116–17, 189
Matilde d', of Tuscany 59
Matilde Bentivoglio d', princess of Modena 10
Rinaldo d', cardinal 22, 26
Rinaldo I d', duke of Modena 3, 4, 5, 12, 22, 27, 110, 117, 119, 120, 187, 203

Fabbri, Paolo 24 n.
Farnese family 12
Farnese, Ranuccio II, duke of Parma 156
Fermi, Anibal 17
Ferrari, Antonio 21–4, 192
Ferrari, Benedetto 14–16, 18, 23–6, 27–39, 58, 110, 125, 155, 188, 192
Ferretti, Basilio 16, 59
Ferretti, Giovanni 16, 59
Ferri, Nicolo Maria 110 n., 116, 187, 196
Ferri Press 6
filippini, *see* Oratorians
Fontanelli Theatre xi, 5, 156
Forni, Lodovico 197
Frangiolli, Ignatio 118
Freschi, Domenico 202
Frignani, Antonio 17, 18, 117, 125

Gabrielli, Domenico 110 n., 117, 196
Gaffi, Bernardo 117, 196, 197
Gandini, Alessandro 205
Gangiura, Alessandro 199
Garbi, Giovanni Francesco 198
Gargieria, Alessandro 187, 198, 199
Garimberti, Andrea 10, 14, 15, 113
Gianettini, Antonio 5, 60, 66, 68, 95, 109–11, 113, 117–20, 121–32, 156, 169, 188, 195, 196–9
Giannini, Domenico 117
Giannini, Giovanni Matteo 115 and n., 196
Gianotti, Antonio 62, 65, 117, 193, 197
Gianturco, Carolyn 40 n., 41 n., 57 n., 205

Giardini, Giovanni Battista 4, 15, 22 n., 23–6, 27–39, 40–57, 58, 60, 62, 65, 69–80, 110, 113, 116–18, 121–32, 133, 169–71, 187, 189, 192–8
Giberti, Francesco 59
Gigli, Giovanni Battista 202, 203
Giovanardi, Nicolò 95
Grancini, Nicolo 59
Grandis, Vincenzo de 6, 58–61, 69–80, 113, 116, 125, 145, 188, 189, 193, 202
Gregory VII, pope 59
Griffin, Julia Ann 21 n.
Grimani, cardinal 119
Grossi, Giovanni Francesco detto Siface 6, 45, 59, 113, 119, 156
Guelph V of Bavaria 59

Henry IV, emperor 59
Hess, Heinz 206
Hill, John Walter 113 n.

India, Sigismondo d', 4
Innocent XI, pope 12, 110, 113

James II, king of England 11, 12, 65, 110, 116–17
Jander, Owen 2 n., 25 n., 206
Janelle, Pierre 7 n.
Jesuits 2, 7, 9, 14, 30
Johnson, Joyce L. 111 n., 114 n., 206
Josquin des Prez 3

Klenz, William 5 n., 206

Lanciani, Flavio Carlo 111, 195, 202
Lassus, Orlandus 15
Latanzio 11
Legrenzi, Giovanni 19–20, 24, 109, 202
Leopold I, emperor 187
Leporatti, Domenico 117, 202
Lodi, Pio 206
Lonati, Carlo Ambrogio 67, 194
Lorandi, Giovanni Alberto 202
Lorenzani, Giovanni Andrea 195
Louis XIV, king of France 11, 12, 65, 116
Lucchi, Marta 41 n.
Luin, Elizabeth J. 23 n., 59 n., 109 n., 202 n., 206
Lulier, Giovanni Lorenzo 114, 117, 196, 197
Lustig, Renzo 113 n.
Luzzaschi, Luzzasco 3

McCrickard, Eleanor 45 n., 206
Magagni, Pietro 203
Malisardi, Gregorio 193
Mantua, duke of 67
Manzini, Giulio Giuseppe 21, 192

Marino, Goffredo 40
Marovaldi, Giovanni 59
Martelli, Simone 202
Martini, Atto 202
Martini, Giovanni Marco 25, 110 n., 116–17, 192, 197, 200 n., 202
Massoni, Giacomo 110 n.
Maylender, Michele 206
Mazarin, Giulio, cardinal 10, 12
Mazzocchi, Virgilio 98
Melani, Alessandro 60, 111, 119, 195, 198
Milano, Ernesto 200 n.
Millanta, Evilmerodach 114, 196, 198
Mischiati, Oscar 62 n., 206
Molza, Ettore, bishop of Modena 21
Monari, Clemente 119, 198
monastery of San Michele in Bosco (Bologna) 19 n.
Monferrato, Natale 109
Monmouth, duke of 12, 65–6
Municipal Theatre (Reggio) 6, 24 n.
Muratori, Lodovico Antonio 206

Nardi, Giulio Cesare 15, 110
Nencini, cavalier 59, 202
Neri, Giovanni Battista 95–108, 194, 198
Nuns of the Madonna, *see* Tertiaries of St Dominic

O'Loughlin, Niall 113 n., 206
Oman, Carola 9 n., 11 n., 206
Oratorians 7, 25, 62, 113, 117, 186
oratory of San Carlo rotondo xi, 2, 6, 8 and n., 9, 18, 21, 22, 25, 27, 30, 40, 58, 60–3, 68, 69, 81, 95, 110, 115, 121, 133, 145, 186, 189, 192–7, 200
oratory of San Geminiano 186, 193, 200 n.
oratory of San Marcello (Rome) 186
oratory of Santi Sebastiano e Rocco (Bologna) 63
oratory of the Santissima Annunciata xi, 187, 197, 200 n.
Order of Clerks Regular, *see* Theatines
Origoni, Marco Antonio 16, 40, 156
Orsi, Astorre 19–20
Orsini, Lelio 193
Ottoboni, Pietro 197, 202
Ottoboni, prince 63 n.

Pacchioni, Antonio Maria 21, 22–3, 25, 59–60, 192, 193
Paini, Giuseppe 14, 16, 18, 58
Paino, Alfonso 18–19, 192
Palazzo Panfilio on the Corso (Rome) 114
Palazzo Riario (Rome) 110
Palermino, padre 62, 64–5, 115–16, 133–44, 146, 171, 193, 196

Palestrina, Giovanni Pierluigi da 63
Pallai, Michele 198
Pallavicino, Carlo 114–15, 146, 196
Paltrineri, Ignazio 22
Panfilio [Pamphilii], Benedetto 62–3, 81, 193–7
Panfilio [Pamphilii] family 62–3
Partenio, Domenico 109
Pasquini, Bernardo 60, 62, 63, 81, 110, 111, 114, 117, 188, 193, 195, 197
Passo, Gioffredo 59
Pastor, Ludwig 8, 206
Perti, Giacomo Antonio 60, 62, 63–4, 114, 188, 193–4
Petrucci, Pier Matteo 202
Pietrogalli, Antonio detto Cottino 16, 156
Pincelli, Paolo 118
Pisani, Sigismondo Gregorio 17, 59
Pistocchi, Francesco Antonio 16, 119, 156–68, 169, 188, 198
Pistocchi, Giovanni 16
Pratichista, Francesco 145

Ricci Press 6
Rizzini, Giovanni 65
Roncaglia, Gino 6 n., 19, 21 n., 58 n., 206
Rosati Press 6
Rosselli, Giovanni Battista 23–4, 192, 197
Rostirolla, Giancarlo 205
Rubino, Nicolo 4

Sacrati, Francesco 194, 196, 197
saints:
    Agnes 63
    Amadeus of Savoy, the Blessed 8
    Antony 21, 71, 111
    Beatrice I d'Este 11
    Beatrice II d'Este 11
    Benedict 26
    Cecilia 19, 63
    Charles Borromeo 9
    Contardo d'Este 8, 11, 21–2
    Francis de Sales 7
    Gaetano 8
    Ignatius of Antioch 22
    Ippolito Galantini, the Blessed 8–9
    Philip Neri 7
    Teodosia 63
    Teresa 7
Salicoli, Margarita 120
Sanuti, Giovanni Battista 25
Sarti, Andrea 117
Sartorio, Filippo Maria 200 n.
Savaro di Mileto, Giovanni 60
Scarlatti, Alessandro 62, 63, 67, 81–94, 156, 167, 188, 194
Schenk, Erich 6 n., 206

Schering, Arnold 46 n., 115 n., 206
Sesini, Ugo 20 n.
Seta, Pietro Paolo 195
Severi, Pellegrino 17, 118
Siface, *see* Grossi, Giovanni Francesco
Sisters of the Visitation of St Francis de Sales,
    see convent, Salesian
Smither, Howard E. 7, 20 n., 66 n., 111 n.,
    113 n., 114 n., 205, 206
Soli, Gusmano 8 n., 9 n., 206
Soliani Press 6, 18, 81, 115 and n., 186, 200 n.
Sossaj, Francesco 2 n., 10 n., 206
Spaccini, Giovanni Battista 3 and n.
Spagna, Arcangelo 82, 145, 157
Spinelli, Alessandro Giuseppe 206
Stampiglia, Silvio *156–68*, 198
Stanzini, Tomaso 195
State archive 14, 40, 45, 74, 115
Stradella, Alessandro 25, 26, *40–57*, 59, 61,
    62, 67, 98, 113, 129, 169, 188, 192, 196,
    202
Stuart family 12
Stuart, James, duke of York, *see* James II, king
    of England
Stuart, Mary, queen of England 116

Tagliavini, Lodovico 4, 15, 115 n.
Tardini, Vincenzo 207
Tasso, Torquato 3
Teatro di Corte 187, 198, 200 n.
Tertiaries of St Dominic xi, 2
Testi, Fulvio 3
Theatines 2, 3, 7–9, 11, 14, 30, 109, 186
Tiraboschi, Girolamo 18, 22 n., 207

Tonani, Pietro 15
Torti, Francesco 197, 199
Trombetta, Agostino 16, 59

Uccellini, Marco 3, 6
university 2
Urio, Antonio Francesco 203

Valdrighi, Luigi Francesco 3 n., 209
Vatielli, Francesco 207
Vecchi, Antonio 196
Vecchi, Orazio 3
Velasquez, Diego Rodriguez 3
Vicini, E. P. 3 n.
Vinacesi [Vinaccesi], Benedetto 95, 119,
    169–85, 188, 198–9
Vitali, Giovanni Battista 3, 5, 6, 16, 18–20,
    25, 58–9, 61, 65–6, 68, 81, 117–19, 125,
    145–55, 188, 194, 197
Vitali, Tomaso Antonio 17, 118
Vitaliani Press 6
Viviani, Giovanni Bonaventura 197

Wasserman 'austriaco' 192
Whitehall Palace (London) 12
Widman, cardinal 15
Willaert, Adriano 3
William of Orange, king of England 116
Worsthorne, Simon Townley 59 n., 207

Zerbini, Pietro 4
Ziani, Marc Antonio 67–8, 194
Ziani, Pietro Andrea 24 n.